T0257835

Natural Gas Technology
Volume II

Natural Gas Technology
Volume II

Edited by **Oliver Haghi**

CLANRYE
INTERNATIONAL

New Jersey

Published by Clanrye International,
55 Van Reypen Street,
Jersey City, NJ 07306, USA
www.clanryeinternational.com

Natural Gas Technology
Volume II
Edited by Oliver Haghi

© 2015 Clanrye International

International Standard Book Number: 978-1-63240-371-1 (Hardback)

Contents

Preface

Over the recent decade, advancements and applications have progressed exponentially. This has led to the increased interest in this field and projects are being conducted to enhance knowledge. The main objective of this book is to present some of the critical challenges and provide insights into possible solutions. This book will answer the varied questions that arise in the field and also provide an increased scope for furthering studies.

Natural gas is one of the cleanest, safest, and most beneficial of all energy sources, and facilitates in meeting the world's increasing needs for cleaner energy for the future. It is an essential constituent of the world's reserve of energy and a crucial basis of numerous bulk chemicals and special chemicals. However, exploring, generating and bringing gas to the user or transforming gas into desired chemicals is a systematical engineering project, and every step needs thorough understanding of gas and the encircling environment. Any developments in the method link could lead to a paradigm shift in gas industry. There have been expanding efforts in gas industry in the last few years. With state-of-the-art inputs by known experts of this field, this book targets the technology developments in natural gas industry. This book is divided into three sections which cover marketing & transportation, utilization and combustion.

I hope that this book, with its visionary approach, will be a valuable addition and will promote interest among readers. Each of the authors has provided their extraordinary competence in their specific fields by providing different perspectives as they come from diverse nations and regions. I thank them for their contributions.

Editor

Part 1

Natural Gas Marketing and Transportation

Natural Gas Market

Joseph Essandoh-Yeddu

University of Cape Coast, Energy Commission,
Ghana

1. Introduction

The natural gas market is the collection of entities or players including buyers and sellers of the gas that compete among themselves or work together on different segments of the gas value and supply chain. It therefore encompasses all the players – producers, transporters, regulators, sellers and buyers and their activities in the market and industry.

Most natural gas trading shares some standard specifications which include the following:

- Specifying the buyer and seller.
- The price.
- The quantity of natural gas to be sold.
- The receipt and delivery point.
- The tenure of contract.
- Payment dates.
- Quality specifications for the gas.
- Arbitration in times of misunderstanding.

Physical contracts are usually negotiated directly between buyers and sellers but electronic bulletin boards and e-commerce trading sites are taking over the transactions in recent times.

Natural gas before the 1970s was once considered a mere byproduct of oil and thus not worth the significant capital investment required to find, gather, process and transport or distribute this resource. So for many places where oil was being prospected and natural gas was found, the wells would be ceiled for locations where local usage demand did not exist.

Driven by energy security and greenhouse gas emissions concerns, the wide spread and cleanliness of the gas has pushed it to the forefront of the fossil fuels such that it has become the clean fossil fuel of choice as well as a global commodity. This has changed the global landscape such that it has become almost equally valuable as the oil being prospected wherever it is found at present times. One important advantage that natural gas has over other fossil fuels besides, its relatively low carbon emissions, is that it leaves no solid residues and produces less other pollutants like sulphides/sulphates on combustion.

Worldwide, the natural gas industry has grown rapidly in recent years. IEA (2011) projects that the global energy share of natural gas would increase from 21% in the 2008-2010 to about 25% by 2035 (Figure 1).

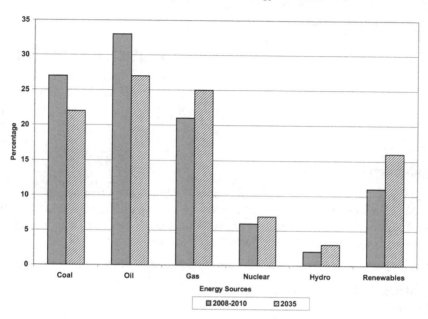

Fig. 1. Global Primary Commercial Energy Demand Shares.

Global marketed gas production has grown from an average of 1.2 trillion cubic metres[1] in the 1970s to about 3.3 trillion cubic metres in 2010. Russia and the United States with annual average production each of about 550 billion cubic metres since 2008 are the largest producers but also the largest consumers, with the United States consuming over 600 billion cubic metres and almost twice the consumption of Russia (BP, 2011).

Global average annual production had almost matched consumption since 2000; 3.05 trillion cubic metres (296 bcf/d)[2] and 3.03 trillion cubic metres (294 bcf/d) respectively. Mean annual growth rate was about 3.2% for both consumption and production.

This significant growth in gas consumption shot up the average price of the gas in the United States from about $6/MMbtu[3] in the 2000s to about $12/MMbtu in early 2008. The increasing price signals accelerated the pace of investment in unconventional gas production facilities, particularly in the United States (IEA, 2011).

The global financial and economic crisis in 2008 however led to a sudden and pronounced drop in consumption from the 3.03 trillion cubic metres in pre-2008 to about 2.94 trillion cubic metres by end of that year. Consumption growth rate between 2007-2008 was 2.4% but dropped to negative 2.2% in 2009. Average annual production of 3.05 trillion cubic metres from 2000-2008 correspondingly dropped to 2.97 trillion cubic metres in 2009.

---

[1] 1.09 billion tonnes of oil equivalent (BTOE) or 116.5 billion cubic feet of gas per day (bcf/d).
[2] Billion cubic feet per day
[3] MMbtu is million British thermal unit

The significant fall in gas demand had led to emergence of significant amount of over capacity in gas production in the United States, causing the average price of the gas in the country to fall to $4.00-$4.50/MMbtu range since the late 2008.

Consumption growth has however risen again; it jumped to 7.5% in 2010 (from negative 2.2% in 2009) dragging production to increase to 3.19 trillion cubic metres (almost 310 bcf/d) that year (from 2.97 trillion cubic metres in 2009).

Most of the world's proved natural gas reserves (about 72%) are located in two regions: the former Soviet Union (FSU) and the Middle East, although there is little production relative to the size of its reserves in the latter region. Both regions hold roughly about 56 trillion cubic metres (about 2,000 trillion cubic feet) of natural gas reserves as compared to the estimated world total reserves of 150 trillion cubic metres (5,300 trillion cubic feet) (CEE, 2006).

Despite the size of their reserves, North American region rather than the former Soviet Union and the Middle East has been the leading producer as well as consumer of the gas for decades. Nevertheless, production in the former Soviet Union region had increased significantly from 0.2 trillion cubic metres (19 bcf/d) in 1970 to 0.71 trillion cubic metres (69 bcf/d) in 2000, i.e. over 260 percent increment whilst the North American production increased only from 0.64 trillion cubic metres (62 bcf/d) to 0.74 trillion cubic metres (72 bcf/d); a mere 16% increase over the same period (CEE, 2006). The former is obviously catching up and this scenario is likely to influence the evolving gas market and industry into a new paradigm in the years to come.

2. Natural gas value chain

Generally, worldwide, natural gas market and industry comprises Upstream, Midstream and Downstream. Their various operational elements are elaborated below (Figure 2).

2.1 Upstream

Upstream activities cover exploration, field production, processing of the raw or associated gas and separation into various other molecules, gathering of the gas from feeder wells before it is pumped midstream.

Production involves extraction of discovered supplies from hydrocarbon fields either with crude oil (as associated/dissolved gas[4]) or separately (non-associated natural gas[5]).

Production time span for gas used for economic analysis is usually longer than oil and can sometimes go up to 30-35 years. During oil production, excess associated gas is disposed of by direct (venting) or by controlled burning (flaring). Under normal conditions, flaring would only take place in the start-up phase of production. Flaring of natural gas however, releases carbon dioxide, the most significant greenhouse gas into the atmosphere. Besides, the flaring gas is also a resource and commodity being wasted since it could be used for power production or as industrial feedstock or exported to earn foreign exchange, of course, all depending upon the quantities involved.

---

[4] Natural gas that occurs in crude oil reservoirs either as free gas (associated) or a gas in solution with crude oil (dissolved gas).
[5] Natural gas that is not in contact with no or significant quantities of crude oil in the reservoir.

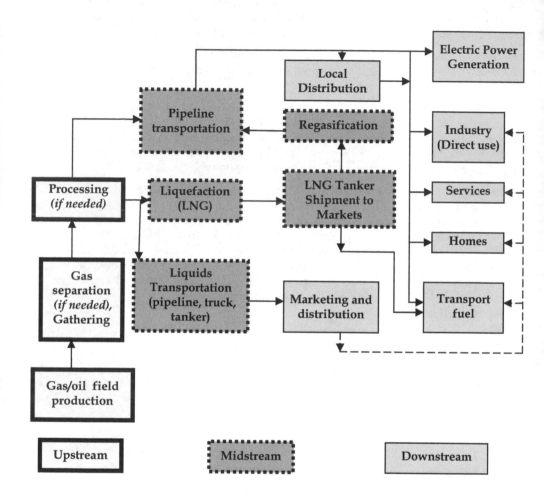

Fig. 2. Natural gas major value chain.

For where technically and economically possible and attractive therefore, the gas is re-injected into the geological formation for latter use and also for secondary recovery to enhance production. Re-injection also is a way of prolonging and increasing or maintaining the oil output. The longer however the re-injection, the likelihood of also losing some significant quantities of the gas in the geological formation, and in some extreme cases, if care is not taken, could damage the reservoir. So re-injection goes with some risk and for that matter gas flaring option technically tends to be more attractive in most cases to the field production operator.

Globally, between 10-15% of the gas produced annually is used for re-injection. Further 5-10% of the gas is flared. Owing to environmental concerns however, pressure to monetize 'flaring' gas is greatly being directed at locations where crude oil is being produced and the associated gas is being flared (IEA, 2011).

In 2010, out of the global production of about 3.2 trillion cubic metres, around 500 billion cubic metres of gas was used for re-injection About 250 billion cubic metres was lost through 'shrinkage' due to extraction of natural gas liquids including LPG[6] and for fuelling production facilities. Between 100-125 billion cubic metres was flared largely in developing countries and in large oil field operations as in Western Siberia of Russia. Much of the flaring however is expected to be reduced significantly by 2020 as most of such host countries are putting in place policies to compel operators to adopt 'zero flaring' of this potential fuel source (IEA, 2011).

Gathering involves collection of natural gas production from multiple wells connected by small diameter pipeline systems to a large pipeline for treatment.

Processing is the separation of heavier molecules and unwanted substances such as water. Out of the heavier molecules could be LPG or Natural Gas Liquids (NGLs) – ethane and larger molecules are stripped out as feedstock for petrochemical industry. LPG, a propane/butane mixture, could be obtained during the processing of **wet** natural gas[7] to obtain the **dry** natural gas[8]. If the gas stream contains impurities such as sulphide compounds then **treatment** is required. The gas is branded as **sour** if it has excessive sulphur compounds, giving it offensive odour. It is **sweet**, if it has much lower concentration of sulphur compounds, particularly, hydrogen sulphide.

2.2 Midstream

Midstream activities include pipeline transportation, liquefaction of the gas into LNG and storage of the gas. Some **upstream** activities however over lap with those of **midstream**. For instance, if the **gathering** involves delivery to a processing plant through a long-distance low pressure pipeline, it is usually placed under **midstream.** Some literature also consider **processing and treatment** of the raw gas, like gas-liquid extraction as a **midstream** activity.

Storage is usually in depleted underground reservoirs or caverns like those associated with salt domes. Storage can be located either near production or near demand. Re-enforced steel

---

[6] LPG is liquefied petroleum gas.

[7] A mixture of hydrocarbon compounds and small quantities of various non-hydrocarbons existing in the gaseous phase or in solution with crude oil in porous rock formations at reservoir conditions. When this mixture gas reaches the surface at normal temperature and pressure conditions, some of the hydrocarbon molecules become liquid. The principal hydrocarbons normally contained in the mixture are methane, ethane, propane, butane and pentane. Typical non-hydrocarbon gases that may be present in reservoir natural gas are water vapour, carbon dioxide, hydrogen sulphide, nitrogen and traces amounts of helium.

[8] Natural gas which remains after; (a) the liquefiable hydrocarbon portion has been removed from the gas stream, (b) any volumes of non-hydrocarbon gases have been removed where they occur in significant quantity to render the gas unmarketable. Dry natural gas is also called consumer-grade natural gas. Dry gas also indicates that the fluid does not contain enough of the heavier molecules to form liquid at surface temperature and pressure conditions.

dome engineered storage systems are also available where natural gas is stored in the form of LNG.

Pipeline transportation covers delivery of the gas from producing basins to local distribution networks and high-volume users via large diameter, high volume pipelines. Countries vary greatly with respect to allowable pipeline specifications for heat content. Some pipelines transport **dry** gas and some **wet** depending upon the standard set by the operators and the buyers.

Liquefaction, shipping and re-gasification known collectively as the LNG value chain, entails conversion of the gas to a liquid form via refrigeration to a cryogenic fluid (temperature -160°C[9]) for transportation from a producing country or region to a consuming country or region via ship.

2.3 Downstream

Midstream activities yield **downstream** activities. **Downstream** covers end-use conversion and distribution of the gas for power generation and to various sectors of the economy.

Distribution **downstream** involves retailing and final delivery of the gas via small diameter and low pressure local gas networks operated by local distribution companies (LDCs) - often called gas utilities. **End use conversion** covers direct use or conversion for use in other forms (petrochemicals, electric power or vehicle fuels).

Overall, there are also losses from **upstream** to **downstream**; for instance, about 20%, i.e. between 800-900 billion cubic metres of the gross gas volume produced globally in 2010 never reach the market (IEA, 2011).

2.4 Commercial and other practices governing the natural gas value chain

The following commercial elements and practices serve to bind the operating segments of the natural gas value chain and link suppliers, transporters and distributors with their customers. They are:

- Aggregation: Consolidation of supply obligations, purchase obligations or both as means of contractually – as opposed to physically – balancing supply and demand.
- Marketing: Purchase of gas supplies from multiple fields and resale to wholesale and retail markets.
- Retail marketing constitutes sales to final end users (typically residential, commercial, industrial, electric power and services).
- Capacity brokering: Trading of unused space on pipelines and in storage facilities.
- Information services: Creation, collection, processing, management and distribution of data related to all the other industry and market functions.
- Financing: Provision of capital funding for facility construction, market development and operation start-up.
- Risk management: Balancing of supply, demand and price risks.

---

[9] -256 °F

- Environmental issues: Impact of the activities of the various segments of the value chain on the ecosystem, comprising land, water bodies, and the atmosphere including the living beings within them.
- Health and Occupational Safety: impact of the various activities on the health and safety of workers in the various segments of the value chain.

2.5 Challenges for efficient and harmonious market along the value chain

The **first** challenge is to create a framework for efficient production of the gas upstream (Foss, 2005).

The **second** challenge is achieving efficiency in pipeline transportation to distribution of the gas to various consuming sectors.

The **third** challenge is developing transparent markets for natural gas supply and consumption. The evolving trend has been to separate infrastructure and product (often referred to as "unbundling") and to search for ways of providing competitive access to pipeline systems for multiple suppliers and users of natural gas (often termed "third party access" or "open access"). In these cases, pipelines become like toll roads, priced through tariff design, while the gas is priced in discrete competitive markets. Traditionally or previously, it was a "bundled" market.

When pipelines become subject to "third party access" regimes, the linking of suppliers to buyers, can be separated into competitive business activities. Such has triggered growth in marketing and trading (both of the physical product as well as financial derivatives) as separate businesses, particularly in North America.

Both the evolution of market-based policies for natural gas and international trade linkages have given rise to a **fourth challenge** which is, timely and accurate data and information on supply, demand and prices. Data and information must be accurate and timely to consistently attract the necessary long-term investment and minimize market disruptions and distortions. Unreliable forecasts create conflicts to the extent that they result in supply-demand imbalances which neither industry nor government at times has the flexibility to correct in a timely manner.

A **fifth** and increasingly **challenge** is dealing with integration, with respect to industry organization and international trade. Industry organization encompasses both vertical (meaning up and down the value chain) and horizontal (meaning over some geographic or international market region) integration. Integration of physical infrastructure across international boundaries has grown rapidly with increased demand for overland pipeline natural gas.

3. Recoverable reserves

Recoverable resources of conventional gas[10] worldwide are estimated to be about 400 trillion cubic metres based on current technology and price range and this is equivalent to more than 120 years of global consumption today (Table 1).

---

[10] i.e. including associated/dissolved gas

| REGION | CONVENTIONAL | | UNCONVENTIONAL | | | | | |
| | | | Tight Gas | | Shale Gas | | CBM | |
	tcm	$/MMBTU	tcm	$/MMBTU	Tcm	$/MMBTU	tcm	$/MMBTU
Eastern Europe and Eurasia	136	2-6	11	3-7	N.A-	N.A	83	3-6
Middle East	116	2-7	9	4-8	14	N.A		N.A
Asia/Pacific	33	4-8	20	4-8	51	N.A	12	3-8
OECD North. America	45	3-9	16	3-7	55	3-7	21	3-8
Latin America	23	3-8	15	3-7	35	N.A	N.A	N.A
Africa	28	3-7	9	-	29	N.A	N.A	N.A
OECD Europe	22	4-9	N.A-	N.A-	16	N.A	N.A	N.A
Total	404	2-9	84	3-8	204	3-7	118	3-8

tcm: trillion cubic metres; MMBTU: million British thermal unit; CBM is coal-bed methane
Source: IEA, 2011

Table 1. Global recoverable resources of conventional gas.

Unconventional resources such as shale gas and coalbed methane (CBM) are now estimated to be as large as conventional resources (Table 1). Production of unconventional gas is estimated to represent about 13% of global production as at 2010 (Table 2).

| REGION | PROVED RESERVES as at 2010 | | PRODUCTION | | CONSUMPTION | | TRADE MOVEMENT | | | |
| | | | | | | | LNG Exports | | Pipeline flows | |
	tcm	%	bcm	%	bcm	%	bcm	%	bcm	%
North America	9.9	5.3	826.1	25.9	846.1	26.7	20.0	6.7	123.6	18.2
S & C America	7.4	4.0	161.2	5.0	147.7	4.7	9.2	3.1	14.3	2.1
Europe & Eurasia	63.1	33.7	1,043.1	32.7	1,137.2	35.9	87.8	29.5	470.0	69.4
Africa	14.7	7.9	209.0	6.5	105.0	3.3	0	0.0	4.9	0.7
Middle East	75.8	40.5	460.7	14.4	365.5	11.5	2.9	1.0	31.5	4.6
Asia – Pacific	16.2	8.7	493.2	15.4	576.6	18.2	177.8	59.7	33.4	4.9
Total	187.1		3,193.3		3,169		297.6		677.6	

NB: tcm is trillion cubic metres; bcm is billion cubic metres
S & C America is South and Central America
Source: BP, 2011

Table 2. Proved Reserves as against Production and Consumption of natural gas in 2010.

Production cost of unconventional gas in North America ranges from $3-7/MMBtu. Other regions could be higher (**Table 1**).

Timely and successful development of new fields however depends on complex factors including industry's capability, policy choices of host countries and market demand.

4. Natural gas demand

IEA (2011) estimates that global gas demand could reach 5.1 trillion cubic metres by 2035 from about 3 trillion cubic metres in 2010. To meet the growing demand, global gas production must increase by 1.8 trillion cubic metres annually and unconventional resources would account for about 40% of it (IEA, 2011).

At the regional level, it seems clear that United States and Canada would continue to play minor role in net global gas trade in the short to medium term due to adequate gas and over-capacity production in North America. On the other hand, rapid growth in Asian gas demand during the same period would put pressure on supply (notably LNG) and consequently, prices and would stimulate greater supply investments within the Asian region and Greenfields elsewhere, particularly in Africa.

All regions are expected to see significant gas growth in the long term (IEA, 2011).

Most important factors determining the demand for natural gas in a country include the following:

- Level of economic activity of the country.
- Richness or suitability of the gas as feedstock for industry.
- Ease and access to supply.
- Competitiveness of the gas versus other fuel sources regarding pricing and environmental considerations.
- Host government policies involving incentives and commercial framework for investors.

Level of economic activity being driven by rising household incomes and increased commercial activities would boost demand for secondary and modern energy such as electricity and gas for cooking and heating. Natural gas use in residential, commercial and transport sectors of the economy requires construction of distribution networks which could be very expensive. Higher household income levels would therefore be necessary to cover cost of delivery and service. Growing industrial sector would require more power to support the growth.

Rich or wet gas if processed could yield other Petroleum Chemicals and for that matter add more value to upstream-midstream activities culminating in industrial growth.

Ease and access to supply of the gas if the price is competitive compared to alternative fuels would attract more customers for the gas. Natural gas has become environmentally **competitive fuel** of choice for environment-conscious economies and also economically competitive for carbon constrained economies because it emits less greenhouse gas per unit combustion compared to other fossil fuels. The global push to utilize relatively clean burning natural gas for power generation has both environmental benefits and power generation diversification and have triggered strong convergence between the natural gas and electric power value chains.

Power generation is the largest gas consuming sector and the biggest driver of gas demand today and it is expected to be so for the foreseeable future. Natural gas is used in both gas turbine and combustion/steam generating plants and prices of the gas in many parts of the world had remained strongly linked to oil, due to its interchangeability with other fossil fuels as industrial and power generation fuel.

Gas demand in the power sector is however said to be more sensitive to changes in the rate of growth of GDP than its use in the other applications. It was observed that from 1990-2008 a global average increase of 1% in GDP led to 1% demand in gas for power generation (IEA, 2011).

Besides power generation, transport stands out as having significant potential to drive demand upwards. Natural gas to supplement or substitute diesel as transportation fuel is one of the measures to reduce urban transport pollution. This is available in a number of European countries but also some developing countries like India.[11]

Natural gas vehicles (NGVs) are fuelled by CNG (compressed natural gas) and LNG (liquefied natural gas). The vehicular features are similar to the conventional ones using internal combustion engine except for the fuel injection system and the size of the fuel storage tank. These differences have given rise to additional costs of $2,000-10,000 higher than the standard vehicles for the natural gas fuelled vehicles.

LNG ships are currently diesel or light crude fuelled but most could be replaced by LNG in the long term and in the foreseeable future. There are approximately 120 LNG ships currently involved in worldwide trade.

Host government policies should be geared towards the ability to facilitate investment in pipeline networks. It means having commercial frameworks in place that not only attract investors but also protect affected public interests. For regional natural gas trade to occur, contiguous countries must have commercial frameworks that are similar enough to encourage market participants to develop efficient cross-border networks as well as dealing with risks. A good example for developing countries is the West African Gas Pipeline (WAGP) that carries Nigerian gas to Benin, Togo and Ghana (WAGPCO, 2011).

5. Natural gas supply

Most of the supply of natural gas is by pipeline for overland delivery and ship for transoceanic delivery as LNG.

Most of the imported gas would come from countries forming the Gas Exporting Countries Forum which has Russia, Oman, Malaysia, Algeria, Nigeria, Equatorial Guinea, Brunei, Iran, Turkmenistan, Indonesia, Qatar and Norway as the main members.

Globally, LNG trade is divided into two international trade blocks; West of Suez Canal and East of Suez Canal. There are old and new exporters as well as new and old importers for both sides of the Suez (Figure 3). For the new, the list is still evolving.

---

[11] Natural gas compressed to about 200 times at standard atmosphere.

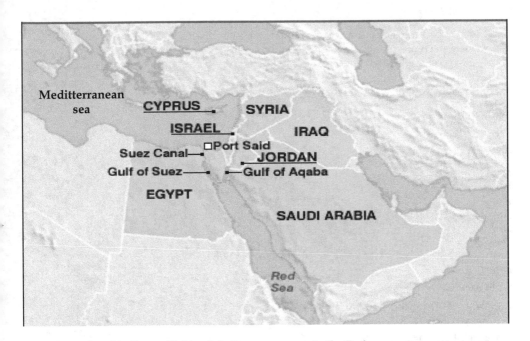

Fig. 3. Suez Canal in Egypt, linking Mediterranean sea to the Red sea.

For the countries involved as at the time of compiling the lists, the picture is as below (Tables 3 and 4):

5.1 East of Suez

Old Exporters	Old Importers
Indonesia	Japan
Malaysia	South Korea
Australia	Taiwan
Brunei	

New Exporters	New Importers (started since 2005)
Qatar	China
UAE	India
Yemen	
	Future
	Thailand
Future	Pakistan
Iran	

Table 3. LNG Exporters and importers east of Suez Canal.

5.2 West of Suez

Substantial LNG regasification was built in the United States early 2000s in anticipation of the country becoming a large importer of LNG. However the unexpected rise in domestic (local) gas production particularly from shale gas during the same period reduced imports to the country significantly.

The following sections would elaborate gas transportation options.

Old Exporters	**Old Importers**
Algeria	Western Europe
Nigeria	USA
Trinidad and Tobago	
New Exporters	**New Importers**
Equatorial Guinea	Western Europe
Egypt	USA
Angola	Brazil
Norway	
Peru	*Future*
	Ghana as hub for West Africa
Future	
North America	

Table 4. LNG Exporters and importers West of Suez Canal.

6. Natural gas transport

The prevailing global market structure is such that for conventional gas, most consumption takes place far away from gas producing centres, which means the gas has to be transported by overland pipelines and or as LNG via specialized ships depending upon the distance and location of the consuming market and this condition is prevalent among developing and emerging market countries that are gas rich. In some cases however, gas is stranded due to physical distances and technical complexities associated with transportation.

Within local or domestic markets, there thus must be sufficient demand and the capacity for potential consumers to pay in order to justify investment in transportation and distribution infrastructure. For most developing countries therefore, the gas is largely used in industry as feedstock or fuel, thus very little goes to the residential and the commercial sectors due to relatively low income per capita. In the developed countries on the other hand, residential, commercial and the industrial sectors compete for the gas almost equally.

6.1 Overland pipeline transportation

Overland pipelines are generally found to have lower cost per unit and higher capacity compared to shipment by rail or road and for that matter the most economical way to transport large quantities of natural gas[12] over land. Overland pipeline transportation and distribution have therefore being the dominant mode for terrestrial gas transport.

The oldest gas pipelines in the world are said to be located in the United States. The first long-distance natural gas trunklines to serve the Midwest Region from the prolific Hugoton Basin located in the Texas/Oklahoma Panhandle were built during the 1930s (U.S EIA, 2011).

---

[12] Also crude oil and refined products.

North America also has the world's longest gas pipeline network. The Canadian system is about 20,000 km whilst the United States has over 90,000 km of natural gas pipeline on more than 66 intrastate natural gas pipeline systems (including offshore-to-onshore and offshore Gulf of Mexico pipelines) delivering gas to local distribution companies, municipalities and to many large industrial and electric power facilities (U.S EIA, 2011).

Europe on the other hand has 28,000-29,000 km of gas pipeline network. Most of the natural gas being used in Europe is imported via pipelines from Russia, Central Asia, the Middle East and North Africa, with imports from Russia accounting for a quarter of the total supply Table 5.

Pipeline	Origin of Gas	Capacity bcf (bcm)*	Completed	Distance Km
Trans-Austria	LNG - Austria	1695 (48)	1960s[13]	380
Trans-Europa	Northsea to Germany	565 (16)[14]	1974	968
Transitgas	Switzerland	NA	1974[15]	293
Frigg U.K	Northsea to Scotland	NA	1977	362
MEGAL	Chech/Autrian to France	777 (22)	1980	1,115
FLAGS	Northsea to U.K	NA	1982	451
Trans-Mediterranean	Tunisia to Italy	1060 (30)	1983	2,475
West-Siberian [16]	Russia	1100 (31)	1984	4,500
STATpipe	Northsea-Norway	671 (19)	1985	890
Fulmar	Northsea to Scotland	160 (4.4)	1986	290
MIDAL	Northsea to Germany	459 (13)	1992	702
NOGAT	Northsea to Netherlands	NA	1992	62
STEGAL	Germany	NA	1992	314
CATS[17]	Northsea to U.K	618 (17.5)	1993	404
Zeepipel-III	Northsea to Belgium	918 (26)	1993	1,416
Rehden-Hamburg	Germany	NA	1994	132
Eorupipe1	Northsea	635 (18)	1995	660
Netra	Northsea to Germany	NA	1995	408
Maghreb-Europe	North Africa	424 (12)	1996	1,620
Yamal-Europe	Russia	1165 (33)	1997	4,196
WEDAL	Germany – Belgium	353 (10)	1998	319
JAGAL	Russia to Germany	847 (24)	1999	338
Europipe11	Norway to Germany	847 (24)	1999	658
Blue Stream	Russia	16 (0.47)	2003	1,207
Green Stream	Libya (North Africa)	283 (8)	2003-2004	520
South Caucasus	Caspian Sea	247 (7)	2006	700
BBL	Netherlands to U.K	668 (19.2)	2006	230
Franpipe	NA	706 (20)	1998	840
Langeled	Norway	2.5 (0.07)	2007	1,200
South Wales	NA	NA	2007	197
Gazela	Russia border to Czech	NA	2011	176
Total				**28,023**

bcf is billion cubic feet; NA is information not available

Table 5. Major existing and operative gas pipelines in Europe.

---

[13] Extension completed in 2007
[14] Upgraded in 1998 and 2009
[15] Upgraded continued up to 2003
[16] Also called Urengoy–Pomary–Uzhgorod or Trans-Siberian pipeline to Europe
[17] Also called Central Area Transmission System

Australia has around 5,000 km gas pipeline network whilst Asia including Middle East has about 15,000 km pipeline network in operation as of 2011.

The longest gas pipeline in sub-Saharan Africa is the West African Gas Pipeline. The 678 km pipeline was completed in 2006 and transports natural gas from Nigeria to Ghana through Benin and Togo (WAPCO, 2011).

6.1.1 New pipeline projects

Relatively few pipelines are under construction and most are still planned as indicated in **Table 6**.

REGION	PIPELINE NAME	DELIVERY POINT	CAPACITY bcm	STATUS	STATE DATE
Russia	Altai	China	30	Planned	2015
	Russian-Asian Pacific	Korea	10	Planned	2015-17
	Nord Stream	N.W. Europe	27.5	Under construction	2011 end
	Nord Stream 2	N.W. Europe	27.5	Planned	2012
	South Stream	S.E. Europe	63	Planned	2015 end
Caspian / Middle East	Nabucco	S.E Europe	26-31	Planned	2017
	ITGI	S.E. Europe	12	Planned	2017
	TAP	Italy	10+10	Planned	2017
	IGAT 9	Europe	37	Planned	2020+
Caspian	CAGP	China	+30	Under construction	2012
	CAGP Expansion	China	+20	Planned	Post CAGP
	TAP1	Pakistan	30	Planned	2015+
Middle East / Turkey	IPI	India	8	Planned	2015+
	Arab Gas Pipeline	Middle East / Turkey	10	Partially built	n.a
Asia Pacific	Myanmar-China	China	12	Under construction	2013
Africa	GALS1	Europe	8	Planned	2015

Start date are as reported by sponsors
CAGP is Central Asian Gas Pipeline; TAPI is Trans-Afghanistan Pipeline; IRI is Iran-Pakistani India; ITGI is Interconnection Turkey Greece Italy; TAP is Trans Adriatic Pipeline; IGAT is Iranian Gas Trunkline; GALSI is Gasdotto-Algeria-Sardegna-Italia. Source: IEA, 2011

Table 6. New pipelines planned and under-construction in the world.

6.2 Liquefied natural gas transportation

Although pipelines can be built under the sea, that process is economically and technically demanding, so the majority of gas at sea is transported by LNG tanker ships which is also seen as flexible compared to pipeline[18].

---

[18] LNG is natural gas that has been compressed 600 times its volume at standard atmospheric conditions (STP) and cooled to an extremely cold temperature (-260° F/ -162.2° C).

Around 75% of the inter-regional global trade is by LNG shipment as compared to pipeline.

LNG ships vary in size from 20,000 cubic metres to over 145,000 cubic metres cargo capacity but the majority of modern vessels are between 125,000 cubic metres and 140,000 cubic metres capacity (58,000 to 65,000 tonnes).

From LNG Liquefaction Plants, ships travel to demand markets to unload the LNG at specially designed terminals where it is pumped from the ship to insulated storage tanks and regasification Plant, where it is warmed up and converted back to gas before being delivered into the local gas pipeline network. Specially designed vehicle trucks are also used to deliver LNG to other storage facilities in different locations.

The liquefaction segment along the value chain is the most expensive and could constitute about 40% of the total LNG value chain cost. The regas segment is the least expensive depending upon the technology opted for (Table 7).

Production	Liquefaction	Shipping	Regasification & Storage
$0.5-1.0/MMBtu	$0.8-1.2/MMBtu	$0.4-1.0/MMBtu	$0.3-0.5/MMBtu
13–27%	22-40%	10-27%	8-15% (15-30% for storage)
Total value chain cost = $2.00-$3.70; with cost escalation = $2.60 - $4.80 between 2006-2010			

Source: Center for Energy Economics, UT-Austin, USA, 2008

Table 7. LNG Value Chain Costs.

LNG delivery from the vessels is accomplished through the following technologies:

i.   Permanent LNG re-gasification plants.
ii.   Floating Re-gasification plants using grounded LNG vessels which have retired from services.
iii. Temporary or stop-gap through "Energy Bridge Re-gasification Vessels" (EBRVs)

Permanent LNG discharge/re-gasification terminal: Development of permanent LNG re-gasification plant requires at least two years if funding for EPC[19] is readily available.

Energy Bridge Regasification Vessels: The energy bridge re-gasification is the LNG technology that delivers the gas in the shortest possible time; i.e. within a year. Energy Bridge Regasification Vessels, or EBRVs™, are purpose-built LNG tankers that incorporate onboard equipment for the vapourisation of LNG and delivery of high pressure natural gas. These vessels load in the same manner as standard LNG tankers at traditional liquefaction terminals, and also retain the flexibility to discharge the gas in two distinct ways. These are through:

• the EBRV's connection with subsea buoy in the hull of the ship; and
• a high pressure gas manifold located in front of the vessel's LNG loading arms.

Floating Re-gasification plants: Average lifetime of most LNG vessels is 25 years. This means LNG vessels built in the 1980s have become less competitive for transport services. Such an LNG ship is retired and reconfigured as floating LNG re-gasification facility.

---

[19] EPC is Engineering, Procurement and Construction.

Typical LNG ship has capacity of 120,000-125,000 liquid cubic metres (lcm). The larger the containment the greater the application for floating storage and regasification applications. Some 59 ships built worldwide before 1983 with containment between 122,000-133,000 lcm are due for retirement.

The LNG market and industry is expanding rapidly. LNG liquefaction capacity was 270 bcm in 2008 increasing to 370 bcm as of 2011 of which Qatar accounts for over 25%. This is expected to expand to 450 bcm in 2015 and doubled by 2020. New projects under construction are due to be on stream from 2014-2016 (Table 8).

| COUNTRY | PLANT | CAPACITY | | START DATE |
		bcm	MTPA	
Algeria	Skikda (rebuild)	6.1	4.5	2013
	Gassi Touil	6.4	4.7	2013
Angola	Angola	7.1	5.2	2012
Australia	Pluto	6.5	4.8	2011
	Gorgon	20.4	15.0	2014
	Gladstone_LNG	10.6	7.8	2014
	Queensland Curtis	11.6	8.5	2015
	Donggi Senoro	2.7	2.0	2014
Papua New Guinea	PNG LNG	9.0	6.6	2014

Note: Start dates are as reported by project sponsors. MTPA is million tonnes per annum. Source: IEA, 2011.

Table 8. Liquefaction plants under construction by country.

500 bcm additional liquefaction capacity is being evaluated for 2015-2020 of which about 75% of it has been earmarked for construction in Australia, Russia, Nigeria and Iran (IEA, 2011).

6.3 Downstream gas delivery to customers

From the transmission pipelines, the gas reaches the consumers through the distribution network. Direct access to gas supply to maintain higher pipeline load factors improves efficiency of operations. However, occasionally, natural gas delivery commitments to downstream customers could be wholly or partially curtailed due to:

- force majeure conditions;
- compressor (station or unit) unavailability;
- line break; and
- supply unavailability for other reasons other than those mentioned above.

For short term (lasting up to 12 hours or 24 hours) "upset" conditions therefore, **linepack** could be used to address the shortfall. The usefulness of the **linepack** would depend on the duration, magnitude and location of this transient condition.

For long term supply disruptions (lasting days, weeks or months), however, storage facilities or an LNG storage system might be required. Depleted oil or gas reservoirs, aquifers or salt caverns are examples of such storage facilities.

If longer (over 6 months) disruptions are anticipated, alternative gas supply network is essential. In general, a mix of overland pipeline and LNG source, with an associated re-gasification unit, for a country or region is recommended.

Because of some of these challenges, domestic or local pipeline transportation of natural gas tends to be characterized by state intervention in various forms to manage market power and to protect the public interest as well.

The next question posed *is at what price is the gas delivered? How is the pricing of the gas computed and what kinds of contracts are made for the supply and delivery of the gas?*

7. Gas pricing, ownership and supply contracts

7.1 Pricing

Natural gas, particularly LNG is generally priced in CIF[20]. Formula 1 is the general pricing formula:

$$\text{Total price} = \text{Gas head price} + \text{Transportation price} + \text{Other prices.} \qquad (1)$$

Where Other prices could be distribution or profit margin, taxes and depreciation.

Some of the gas pricing mechanisms that have been introduced into the market are as follows:

- *Cost of Service*
- *Spot Price*
- *Netback Pricing*
- *"S-shaped" price curve*
- *Barter trade*
- *Price-Index provision*
- *Seasonal and weather-normalised rates*
- *Gas Swaps*
- *Production or Supply Payment*

7.1.1 Cost of service

The cost of service (COS)[21] of a project is defined as the minimum price required to provide for capital recovery, covering operating costs and paying taxes, royalties and production sharing, etc. The traditional gas market is based on long term contract and has used COS for pricing. COS is also basis for price regulation in most countries including approving appropriate return on investments.

---

[20] Unlike oil which is generally priced in FOB. CIF is abbreviation for Cost, Insurance and Freight, whilst FOB is abbreviation for Free on Board.
[21] Also called "Rate of Return"

The regulated company's revenue requirements are the total funds that the company may collect from customers and it is calculated by multiplying the company's rate base by an allowed rate of return (ROR) and adding this product to the company's operating costs (OC) as shown in formula 2.

$$\text{Revenue Requirement} = (\text{Rate base} \times \text{ROR}) + \text{OC} - (\text{Taxes} + \text{Depreciation}) \quad (2)$$

The Rate base is the total value of the company's capital investments, which may include constructional work in progress. The allowed ROR constitutes a profit sufficient to pay interest on accumulated debt and to provide a negotiated acceptable return to investors.

A negotiated acceptable or what is termed as a fair return is determined through a comparable earning tests, where a company's earnings are measured against those of a firm facing comparable risks[22]. Operating costs include expenses on gas, labour, management, maintenance and commercials. Taxes and depreciation are also part of the company's revenue requirments.

Regulators are faced with the challenges of determining the price appropriate and acceptable to the seller to cover cost operations and future investments. Also, there is a challenge in allocating costs and acceptable prices/tariffs to the different customer classes. The regulator therefore first seeks to determine how much of an applicant's captial stock should be included in its Rate base, then attempts to determine which elements of test year costs and revenues should be allowed for regulatory purposes and whether to allow specific changes since the test year. The final step is to determine what the fair ROR is for the company.

7.1.2 Spot price

Spot price is the price of a commodity such as the gas on the spot and is dependent upon time and location. The spot price of gas say in Takoradi, Ghana at 01.00 GMT could be different from the spot price of gas of the same quality in Takoradi, Ghana at 02.00 GMT.

Spot market is characterized by:

1. Short purchase contract term, usually 18 months.
2. Best-effort delivery.
3. Negotiated price reflecting current market conditions.
4. Arranging transportation separately from the sales and usually provided on an interruptible basis.

There is an increasing frequency of spot sales. Participants in the **spot market** include:

1. Local distribution companies (LDCs) who buy the gas from the pipeline companies.
2. Marketers/Traders who buy the gas on the **spot market** to resell to LDCs and large customers. Some marketers do not sell gas themselves. Instead, they bring buyers and sellers together and help negotiate arrangements for transportation of the gas.

---

[22] A discounted cash flow approach, where a company's capital costs are estimated by analyzing conditions in the financial market, or other methods.

3. Pipeline operators at times purchase **spot market** gas and mix it in their supply portfolios in an effort to lower weighted average cost of their gas.
4. Industry, large commercial consumers and electric utilities may purchase **spot market** gas to increase supply security, but they have to arrange transportation of the gas to their vicinities.

As spot sales increase and proliferate, they are likely to undermine, if not topple the long term contract and price structure which had been an important feature of the traditional gas market.

7.1.3 Netback pricing

Netback pricing is retail price less all costs and expenses. The marketer after taking care of all associated costs, expenses and agreed profit margin, returns to the producer the balance. It encourages the marketer and the producer to increase market share and also receive a fixed profit margin.

Netback Pricing was first introduced by Saudi Arabia in 1985 and it was a revolution because the setting of the pricing was shifted from the producer to the consumer and it was the case until early 2000s when FUTURES market took over.

7.1.4 "S-shaped" price curve

This somehow operates on the principle of price ceiling and price floor. At below an agreed price floor of say $6 per MMBtu, the buyer agrees to pay an additional premium for the gas. However above price ceiling of say $10 per MMBtu, it is the reverse, the seller pays a premium to the buyer. The end result is a win-win situation where the buyer enjoys a discount at high prices and the seller is protected against low gas prices. The **S-shaped** pricing mechanism is likely to be used more in the future, particularly as civil society voice becomes louder to see fair share of profit for particularly host developing countries with the finite resource.

7.1.5 Barter trade

Barter trade is simply the exchange of gas for other commodities needed by the gas-exporting country. Existing barter agreements include Russian gas for Polish potatoes and Russian gas for Ukrainian consumer goods food and machinery.

7.1.6 Price-Index provision

Besides the base pricing described above, there is also the **indexing**. Meaning the pricing formula usually consists of two main parts, the base price and indexing. The major consideration for pricing in the case of gas is that is largely used as fuel and so the price is on energy basis and so indexed to price movement of other alternative fuels, such as crude oil, gasoline and fuel oil. Price-index provision ensures that the gas delivered is competitively priced compared to alternate fuels.

Besides price indexing and to cope with competition from alternative fuels and to avoid losing customers, some pipeline operators would provide special tariffs to keep multi-or dual fuel capable industrial customers from switching during period of high gas prices. They could offer incentives such as waiving transportation rate on the gas to reduce the tariff.

7.1.7 Seasonal and weather-normalised rates

Seasonal rates are where the price of natural gas is influenced by the seasonality of gas demand and supply. It is one of the new pricing mechanism for gas particularly in temperate regions. For weather-normalised rates, the price of natural gas is locked into an agreed weather and this can help reduce the effect of abnormal weather patterns on utility earnings.

Tariffs charged by pipelines and LDCs reflect gas sales and transportation volumes. The higher the volumes, the better the sales. However, residential demand is largely weather-sensitive in cold/ temperate climates. Meaning the warmer the weather the less the demand and vice versa. Abnormal weather patterns have therefore been the single most important factor for supplies to largely residential customers. To minimize the impact of weather on revenues, some gas utilities use the **weather-normalised rate** for customer billing; rates are increased when weather conditions are warmer to cover drop in demand; conversely, rates are reduced when weather is colder than normal.

7.1.8 Gas swaps

Besides, **weather-normalised rate**, an LDC may enter into agreements with a large industrial customer with dual-fuel capacity and high-load factors. In the summer when demand for gas is lower for the LDC, the industrial customers may utilize the LDC's excess supplies. Conversely, in winter periods when demand for gas is higher from the LDC, the industrial customer switch to their alternative fuels to enable the LDC meets their contracted volume supplies. Such **gas swap** contracts result in savings for both the LDCs and the customers.

7.1.9 Production or supply payment

In production or supply payment, the buyer makes upfront cash payment for gas to be produced or supplied over a long term in most cases, years. The major features of production payment are:

1. It represents a new capital source for the gas industry.
2. It is basically risk-free to producers. Further, the buyer only has recourse to the given field specified in the agreement.
3. The buyer is better off locking in firm title to reserves at a known price to back their sale commitments to guide against period of rising prices.

7.2 Ownership contracts

The different types of ownership or operational contracts regulating the gas market are:

• Concession

- Production Sharing Contract or Agreement, i.e. PSC or PSA
- Joint Venture
- Service Contract

Concession

Concession is an agreement that is royalty based. It means the host country government gets most of its share through royalty. Royalty is another name for Production Tax and it is related to gross revenue. Gross revenue less royalty equals NET revenue. Royalty also means, the government of host country gets its pay first. The advantage is that host country or government's risk is zero, because production loss or profit, the government takes the royalty. The next deductions from the net revenue include operating costs, depreciation, amortization and intangible drilling costs. Revenue remaining after royalty and the deductions is called **taxable income**. The remaining revenue after taxation is the **contractors take.**

Production – Sharing Contracts (PSC)

For purely Production Sharing Contract (PSC) or Agreement (PSA) as sometimes called, the operating company takes all the risks but manages the production, etc. Government in turn may take off taxes on all imported equipment and provide other tax incentives. After all the deductions by the operating company, there remains the NET revenue and this is what is shared between the operating company or entity and the government. The net revenue is called the profit hydrocarbon and in this case the **profit gas** and is split between the contractor and the government, according to the terms of the PSC negotiated. It means the company takes its money first before government/host country comes in, whilst in **concession** the government takes its money first.

The title of the hydrocarbons however, remains with the host country government (Johnston, D. 1994). With growing awareness of good governance, the state may also maintain the management control and, or would require the operator to submit annual work programmes for scrutiny and approval. Most PSC/PSAs are placing limit on cost recovery such that if the deductions amounts to more than the allowed limit, the balance would be carried forward and recovered later. In some cases, the host country would push royalty payment into the PSC/PSA, and may go by different names such as War Tax levy in Columbia (Johnston, D. 1994). In this case, the profit gas is Net Recovery less Cost Recovery. The operator's share of the profit gas may also be subject to taxation.

Joint venture

Joint Venture is a partnership where the parties share the risks and the rewards together. In **joint venture,** there is always a reference to WI (working interest) meaning – sharing of all costs and expenditures.

Investment in the gas industry in general is capital intensive and has long lead time and often involves financial risks. Investment through Joint Ventures therefore spreads the risks among parties and therefore reduces the share of responsibility of individual parties involved.

Service contract

Service contract means producing and may be selling in this case the gas on contract for the host country or government. In a typical **service contract** therefore, all the gas belongs to the host country or government. The operating company is paid for every per volume or energy of gas produced.

A kind of service contract is **Technical Assistance Contract** (TAC). TACs are often referred to as rehabilitation, redevelopment or enhanced oil recovery projects. They are associated with existing fields of production and sometimes, but to a lesser extent, abandoned fields. The contractor takes over operations including equipment and personnel if applicable. The assistance that includes capital provided by the contractor is principally based on special know-how such as steam or water flood expertise.

Rate of return / R- factor contract

Some countries have developed progressive taxes or sharing arrangements based on project rate of return (ROR). The effective government-take increases as the project ROR increases. The sliding-scale taxes and other attempts at flexibility may be based on profitability and production rates depending upon negotiation. To be truly progressive however, it should be based upon profitability.

Some contracts use what is called an R factor. **R** factors deal with all variables that affect project economics and it is expressed as:

$$Rfactor = \frac{Accrued.net.earnings}{Accrued.total.\exp enditures} \tag{3}$$

7.2.1 Some basic elements of operating contracts

The existing market is apparently being driven by a hybrid of **concession** and **PSC/PSA**.

PSC/PSA in the long term becomes **Service** contract, provided no new major investment is made in the production business, otherwise in most cases equipment purchased or imported under the contract become the property of the state at the end of the project. This is because, under most PSC/PSAs the contractor cedes ownership rights to the government for equipment, platforms, pipelines and facilities upon commissioning or startup. The government as owner is theoretically responsible for the cost of abandonment. Anticipated cost of abandonment is accumulated through a sinking fund that matures at the time of abandonment. The costs are recovered prior to abandonment so that funds are available when needed.

PSC/PSA is also being phased out in preference for **Joint Venture** contracts and that is the likely contract for the gas industry and market in the future, because host or producing (usually developing) nations demand quick cash revenue turnovers and these are assured under Joint-ventures. Most national oil companies (NOCs) are therefore moving into **Joint – Ventures**.

In summary, the characteristics of the contract types are as below (Table 9):

Contract type	Characteristic
Concession	Royalty for Government first.
PSC/PSA	Share of profit gas but company takes money first.
Joint Venture	Net Revenue interest.
Service Contract	All gas belongs to Government.
Rate of Return	Rate of return

Table 9. Summary characteristics of ownership contracts.

Many aspects of government/contractor relationship may be negotiated but some are normally determined by legislation. Elements not determined by legislation must be negotiated. Even though, it is usually better to have more aspects that are subject to negotiation, flexibility is the watchword.

Elements like **royalty is** taken right off the gross revenues and therefore contributes to their lack of popularity with the industry since they can cause production to become uneconomic prematurely. This works to the disadvantage of both the industry and government. One remedy that has become popular is to scale royalties and other fiscal elements to accommodate marginal situations. The most common approach is an incremental sliding scale based on average daily production. A sample sliding scale royalty could be as below (Table 10).

Average	Daily production	Royalty
First tranche	Up to 100 mmscf/d	5%
Second tranche	101-200 mmscf/d	10%
Third tranche	Above 200 mmscf/d	15%

Table 10. Sample sliding scale royalty.

Cost recovery may include the following items:

- *Tangible and intangible capital costs.*
- *Interest on financing (usually with limitations).*
- *Sunk costs*
- *General and Administrative Cost.*
- *Investment Credits and Uplifts*

Tangible vs. intangible capital costs: Sometimes a distinction is made between depreciation of fixed capital assets and amortization of intangible capital costs. Under some concession agreements, intangible exploration and development costs are not amortized. They are expensed in the year they are incurred and treated as ordinary operating expenses. Instances where intangible capital costs are written off immediately can be an important financial incentive. Amortizing intangible costs can take longer to recover, if not carefully negotiated.

Interest cost recovery: Sometimes interest expense is allowed as a deduction. Some contracts limit the amount of interest expense by using a theoretical capitalization structure such as a maximum 70% debt (Derman, A. and Johnston, D. 1999).

General and administrative costs: Many contracts allow the contractor to recover some office administrative and overhead expenses. Non operators are normally not allowed to recover such costs. Most unrecovered costs are carried forward and are available for recovery in subsequent periods. The same is true for unused deductions.

Sunk cost is applied to past costs that have not been recovered. Exploration sunk costs can have a significant impact on field development economics and can strongly affect the development decision. For this reason, many contracts may not allow pre-production costs to begin depreciation or amortization prior to the beginning of production.

Investment credits and uplifts allow the contractor to recover an additional percentage of capital costs through cost recovery. For example, an uplift of 20% on capital expenditures of $100 million would allow the contractor to recover $120 million. Uplifts can create incentives for the industry. Uplifts are the key of rate of return contracts.

Most contracts have a limit to the amount of revenues the contractor may claim for cost recovery but would allow unrecovered costs to be carried forward and recovered in succeeding years. In summary, the hierarchy of cost recovery can make a difference in cash flow calculations.

The basic elements of operating contracts besides royalty and cost recovery include work commitment, bonus payments, domestic obligation, ring fencing, commerciality, reinvestment obligations, tax and royalty holidays.

Work commitment refers to the obligations an exploration company incurs once a PSC/PSA is formalized and they are generally measured in kilometres of seismic data and the number of wells to be drilled in the exploration phase.

Cash bonuses are lump sums paid by the contractor to acquire a particular license. These cash bonuses are the main element in bidding rounds of very prospective acreage. **Production bonuses** are paid when production from a given contract area or field reaches a specified level.

Domestic obligation: Many contracts specify that a certain percentage of the contractor's profit oil be sold to the government. The sales price to the government is usually at a discount to world prices.

Ring fencing: Ordinarily all costs associated with a given block or license must be recovered from revenues generated within that block. The block is *ring fenced*. This element of a contract can have a huge impact on the recovery costs of exploration and development. From the government perspective, any consideration for costs to cross a ring fence means that the government may in effect subsidize unsuccessful operations. Allowing exploration costs to *cross the fence* may therefore be negotiated.

Commerciality deals with who determines whether or not a discovery is economically feasible and should be developed. Some regimes allow the contractor to decide whether or not to commence development operations. Other systems have a commerciality requirement where the contractor has to prove that the development of a discovery is economically beneficial for both the contractor and the government. The benchmark for obtaining commercial status for a discovery cannot be developed unless it is granted commercial status by the host government. The grant of the commercial status marks the end of the exploration phase and the beginning of the development phase of a contract.

Reinvestment obligations: Some contracts require the contractor to set aside a specified percentage of income for further exploratory work within the license.

Tax and royalty holidays: The purpose of tax and royalty holidays by the host country is to attract additional investment.

7.3 Supply contracts

There are four main types of physical trading supply contracts, namely, **swing, base-load, firm** and **futures contracts.**

Swing contract

Swing contracts are usually short term contracts and not longer than a month. It is also called 'interruptible'contracts. Under this type of contract, both the buyer and the seller agree that delivery of the gas can be interrupted on short notice; no legal commitment. They are the most flexible and are usually put in place when either the supply of gas from the seller, or demand for gas from the buyer are unreliable.

Base-load contract

Base-load contracts are similar to swing contracts in that neither the buyer nor seller is obligated to deliver or receive the exact quantities specified. However, it is agreed that both parties would attempt to deliver or receive the specified volume on a best-efforts basis. In addition, both parties generally agree not to terminate the agreement due to market price movements. There is however no legal recourse for either party if they believe the other party did not make the best effort to fulfill agreement. Such contracts rely instead on the relationship (being it personal or professional) between the buyer and the seller.

Firm contract

Firm contracts are different from swing and base-load contracts in that there is legal recourse available to either party, should the other party fails to meet its obligations under the agreement. These contracts are used primarily when both the supply and demand for the specified quantity of the gas are not likely to change.

Futures contract

Futures is one of the derivatives used in the financial markets for both commodities and securities.

Futures contract entitles the buyer of the gas through the contract to take delivery of it at an agreed location and at an agreed date specified in the contract in the future and it compels the seller to deliver the commodity to the buyer at the specified date in the future under the same conditions. Because the contract is tradable, i.e. can be bought and sold in open market, its value changes as the supply of and demand of these contracts changes.

A common feature for futures contracts is that they are standardized such that each futures contract represents the same quantity and quality, valued in the same pricing format, to be delivered and received at the same agreed delivery location and date. The only variable in a futures contract as to when it is bought and sold is *the price of the contract.*

The success of the natural gas market depends on factors including uncertainty in supply and demand, large trading volumes and price volatility. For this reason, futures traders base their price offers on the spot market prices and the markets also allow companies to hedge their price risks.

The major delivery locations in the United States are the Henry Hub in Louisiana and Waha Hub in West Texas. The high volume of trading activity and the high degree of volatility in prices at these two locations have made them points of choice for traders.

The New York Mercantile Exchange (NYMEX) introduced and began trading in natural gas futures with Henry Hub in 1990. The Kansas City Board of Trade (KSBT) also began trading in natural gas futures in 1995 with Waha Hub as the delivery point.

The International Petroleum Exchange (IPE) opened its gas futures for trading in 1997. More companies are now trading their gas futures through IPE. IPE quoted prices are increasing being used as guide by a number of firms in Europe and are gradually becoming a benchmark for gas market in Europe. IPE operates a 12-month range gas futures contract and it is likely to be extended to 15 months.

8. Gas market and industry structure

8.1 Traditional regulation

In the traditional regulatory environment, the main gas transmission entities have mostly been state owned. The state entity supplies gas to one or more distribution entities that are charged with the distribution of the gas to retail customers in specific concessional areas exclusively. In some cases, the transmission monopoly is legally separate from the distribution entity. Supply contracts between the transmission monopoly and the distribution entities are usually long term and are regulated by government agency. A distribution entity in turn has legal monopoly over the supply to retail customers in a concessional area. The gas supply to the public is also regulated by a government regulatory agency and it covers regulation of prices, return on investment, etc. Some regulatory agencies are fully or partially decentralized to the state, regional or district/local level.

The traditional gas marketing system also bound producers, pipelines, local distribution companies (LDCs) and consumers together with long-term contracts, with little room to respond to changes in the market place.

The government controlled price regulation and political pressures on price levels sometimes lead to operational inefficiencies. As demand grows however, pressure to remove subsidies would increase.

Bundled service

The traditional gas market is characterized by **bundled service** – production, transmission and distribution vertically connected and owned by one entity or a consortium. **Bundled service** certainly gives more monopoly to one company, but provides less complication for a new or emerging industry in a country.

8.2 Unbundling service: The new and future regulation

Unbundling is the process of separating natural gas services and supply into components with each component priced separately. Natural gas companies go through varying degrees of organizational unbundling based on market maturity, monopolistic power of the incumbent and the regulatory regime in place.

Unbundled service is usually appropriate for older, matured markets, typically with hundreds of customers. It is not uncommon to amend regulations to cater for unbundled service as the market gains maturity and more new customers (particularly the commercial and residential customers) come on stream. In the United States for instance, unbundling of the market and transportation services for interstate pipelines did not occur till 1993 (FERC Order 636), even though the networks had been in place for over 50 years. In South America, unbundling was born out of increased consumer groups and industry maturity.

The main objective is to prevent cross subsidy, abuse of market and rather maximize efficiency. The purpose of unbundling is to secure non-discriminatory treatment for companies seeking access to pipeline, by ensuring that a vertically integrated transport company does not discriminate in favour of its own gas supply business.

Unbundling ensures that cost is correctly allocated to transportation. This cost clarity provides a basis for establishing use-of-system charges. Unbundling enables customers to pick and create their own services package. Marketers can package the variety of options and sell these services to consumers without discrimination. This leads to the facilitation of competition, also allowing which components of the value chain could be offered for privatization should the need arise and consequently increased efficiency.

Unbundling would characterize the new and future gas market structure.

8.3 Privatization of the gas industry

Privatization allows governments to attract private capital and encourage private investments in its business portfolios. Objectives are to:

- Restructuring poorly run state-owned entities.
- Raising cash to relieve budgetary deficits.
- Raising foreign capital to repay foreign debt.
- Spreading ownership of operations.

Other advantages include attracting new technologies, increased competition and improved efficiency in operations.

8.3.1 Methods of privatization

i. Public offering of shares

Under this method, the government sells to the general public all or part of the shares it holds in a state-owned company. In both cases, widespread shareholding of the entity or enterprise is created.
When only a portion of the shares is sold, the result is a joint state-private ownership of the entity, or what is currently termed as Public Private Partnership.

ii. Private sale of shares

Under this method, the government sells all or part of its holdings in a state-owned company to a single purchaser or group of purchasers. The transaction can be direct acquisition or through a third party such as a broker.

iii. Asset acquisition

The transaction comprises sale of assets instead of shares. Assets can be sold individually to downsize the entity or, sold bundled to form a new corporate entity. The sale of asset can be by open competitive bidding or direct negotiation with the purchaser.

iv. Fragmentation

Fragmentation involves the breaking up or re-organisation of a state-owned entity into several separate entities or into a holding company with several subsidiaries. This method permits piecemeal privatization and allows other different methods of privatization to be applied to different component parts, thereby potentially maximizing the benefits of the overall process. Fragmentation also allows large state-owned entities into separate enterprises and eventually creating competition in the market.

v. Expanding state owned entities with private investment

Under this method, the government instead of disposing of any of its equity rather invites the private sector to buy into the venture. As with Public offering of shares, it results in a Private – Public Partnership joint venture.

vi. Management/Employee buy-out

Under this method, a group of managers or employees acquires a controlling interest in the entity or enterprise. The management/employee leverage use credit to finance the acquisition whilst the collective assets of the acquired enterprise is used as collateral.

This method provides means of transferring the ownership to management and employees and could be a solution for state-owned companies that are difficult to sell or very strategic to the economy, community or the nation. A strong cash-flow potential however is usually the pre-requisite for securing credit for the buy-out.

vii. Management contracts

Under this method, functions related to the entity's operations are contracted out to external usually private management group. There is usually no transfer of ownership and no divestiture of state assets. It has the potential to increase efficiency and effective use of state assets and it is seen as one of the feasible options for introducing private equity into state-owned enterprise especially in developing countries.

Examples of countries with privatization of gas sector

Argentina in 1985 embarked on privatization drive that led to the sale of its national gas transmission systems and a state-owned gas distribution company.

Belgium deregulated its gas sector in the year 2000 and allowed private participation in its local gas market.

Russian gas industry is dominated by GAZPROM, a state-owned company and it is known to be the largest in the world. In 1993, GAZPROM became a state-owned joint stock company and began privatization in 1994. Management/employees as well as local investors were allowed to buy shares whilst the government retained 40% of the shares. 9% of GAZPROM shares were set aside for foreign ownership.

9. The future market

In the evolving and future market and to the customer, cost of the gas and cost of transportation are the major considerations to determine the least-cost gas supply plan. This new marketing paradigm is putting a lot of pressure on gas producers, pipeline operators and LDCs to compete with each other eventually improving efficiency and potentially leading to decrease in cost of supply. New industry players including spot gas marketers and brokers of pipeline capacity create additional links between suppliers and customers. These developments are expected to intensify competition in the gas market and would have major implications for the industry in the future.

The focus of restructuring and regulatory reform in the evolving market therefore would be to reduce state regulations and introduce more competitive market whilst concurrently reforming the regulations to induce more efficiency in performance.

From the on-going restructuring worldwide, six lessons could be deduced which could also serve as guidelines for restructuring of the traditional gas industry. They are as follows:

i.   Privatizing state-owned entities with the objective to create efficiency;
ii. Promoting competition in the supply of gas services by
     a. Opening up access to new suppliers; and
     b. Deregulating prices
iii. Ensuring that transmission access rules and associated prices are in most cases non-discriminatory to all applicants to support competition, except for foundation customers and where in special cases targeted at promoting local industry.
iv. Developing pricing arrangements that provide revenues to expand efficient investments maintain decent return on investments.
v.   Providing sustainable return on investment for distribution companies but deemed fairly affordable for consumers.
vi. Putting in place regulatory and contractual mechanisms that ensure that market agreements are honoured.

9.1 Mergers and acquisitions in the gas industry

Merger in simple term means legally combining strengths but minimizing weaknesses. The end results include extending cooperative life. Acquisition simply means the entity being acquired has conceded weakness on its part.

Combining two corporate cultures could however cause serious challenges and could lead to failure if not managed well. Fundamental considerations for mergers and acquisition should therefore be:

1. Relate to the core business and that there is expertise to run the expanded business.
2. Supplement or extend current operations.
3. Not raise end-user tariff significantly, in most cases, it should lead to reduction in cost of doing business and eventually decrease end-user tariff.

9.1.1 Targets for mergers and acquisitions

Most attractive targets and successful ventures in the gas industry have been

• Parties involved are in a closely related business.

- The parties involved complement each other's weaknesses.
- Financing does not create unreasonably high debt-to-equity ratio.
- Employees of both parties receive a fair deal and most are happy with the merger or acquisition.

For natural gas pipelines and local distribution companies (LDCs), expansion through mergers and acquisitions offer numerous benefits including:

i.   Reducing over all management costs.
ii.   Creating large customer base.
iii. Opening access to new supply sources.
iv.   Increase economies of scale.
v.   Penetrating new markets and offering new services.
vi.   Reducing or avoiding new investments by gaining access to new facilities.
vii. Cutting costs by eliminating duplicate services.
viii. Establishing name and recognition with customers.
ix.   Flexibility in transporting large volumes of gas due to increase market share.
x.   Expanded operational areas.

Unsuccessful mergers and acquisition on the other hand are characterized by the fact that the acquiring party:

- Pays too much for the acquisition. *Over-value of the acquired assets can cause acquisition failures. This can come about when the evaluator mistakenly over-values the entity being acquired.*
- Over estimate the market projections.
- Combine two different corporate cultures.
- Had limited access to information before merger or acquisition, in other words, lack of transparency.

9.2 National gas and international gas companies

National gas companies

National gas companies (NGCs) are currently and generally the same as the national oil companies (NOCs) of the host countries, since the gas market is still evolving compared to the oil market. NOCs usually represent the interest of their governments in the petroleum market. They act as gatekeepers and control access to the majority of resources for future oil and or gas production.

Many NOCs came into being during a period of relatively large-scale state intervention in their countries' economies, a process which only began to reverse in the 1980s-1990s (Stevens, 2003). It was envisaged in those times that market forces would not be sufficient to propel poor developing host countries out of their poverties. With most natural resources vested in the government, it was thought that only the state could marshall the resources required for the massive economic development. In recent times however, additional reasons have been (i) emergence of nationalism; (ii) hydrocarbon listed as a strategic resource; and (iii) commercially risky and technologically complex sector (Foss, 2005a; Foss, 2005b; Mommer, 2002).

The overall goal of an NOC therefore is to ensure the effective development of the hydrocarbon sector of the country and as well contribute to the country's socio-economic development.

The immediate objectives include:

- Earning revenue for the country;
- attracting new technology;
- ensuring infrastructure development;
- creating employment; and
- the latest addition, minimising damage to the environment.

NOCs differ from country to country based on the following:

- the level of government or state control or ownership.
- The extent of private shares.
- Their financial health.
- Access to international credit.
- Their operational experience and skills.

The more experience, skillful and higher the access to international credit, the more likely such an NOC would go international and become an IOC (International Oil Company).

International oil companies

The traditional examples of IOCs are Exxon-Mobil and Chevron of the United States, BP of United Kingdom, Royal Shell of the Netherlands, Elf and Total of France. The new entries include the PetroChina of China, Petrobras of Brazil, Petromas of Malaysia and Norsk Hydro of Norway.

The primary objectives of an IOC in a host country are:

- To look for good geology to find the resource, in this case gas.
- To expand operations.
- To maximize return on equity to shareholders, simply saying to make money.

What most IOCs consider before entering a host country after the presence of a good geology has been confirmed, include:

- Political stability.
- Respect for honouring contractual obligations, i.e. contract is not unilaterally changed in the middle of the course.
- Robust and transparent legal system, i.e. company believes that it shall receive a fair hearing during legal cases.
- Fulfilling bilateral cooperation between partners in the business and the host country.

Cooperate social responsibility goals have since 1990s been playing greater role in their host country operations.

9.3 Future of national and international gas/oil companies

More NOCs would move away from just exporting raw gas to adding value to the commodity. Nations with rich natural gas resources have aggressively added new petrochemicals capacity for the production of methanol and other industrial chemicals. It has been observed that a country like Trinidad and Tobago has been very successful in attracting foreign investments. In the mid-1970s, the country had a paradigm shift with respect to the focus of the hydrocarbon production – monetization of natural gas. Since 1975, the natural gas has been used to manufacture methanol and ammonia and exported to the United States.

A shift from "raw gas exporting" to "value-added" however would depend upon host Government policies and the type of instruments used to implement the policies, since they have impact on local gas consumption and fuel choices, directly or indirectly. For instance environmental policies to promote cleaner alternative fuels may encourage greater gas use locally if available through favourable taxation and financial incentives for development of infrastructure. Lower taxes on gas prices and CNG vehicles but high on diesel fuelled vehicle could help reduce local consumption of diesel and consequently local pollution in mega cities such as found in India.

With time and as nationalism sentiments grow, traditional IOCs such a Exxon-Mobil would be pushed out of most host developing countries whilst more NOCs would become IOCs. For instance, Petrobras of Brazil, Petromax of Malaysia and PetroChina of China which were formally NOCs have become the major new IOCs.

For IOCs therefore to exist in future, **first**, they must be at the cutting edge of technology where the NOCs are nowhere near. This can only be achieved through research and development (R&D). R&D investment however does not come cheap and does not yield immediate results; it may take between five to ten years.

IOCs that are short-sighted would not invest in R&D and by this posture, would be those which would cease to exist.

As oil and gas resources are becoming dear to explore and produce, the NOCs would look for IOCs with the requisite technologies. IOCs in the future therefore would be looked at as 'banks' of technologies and the knowhow. Therefore robust IOCs shall be the 'banks' with the skills.

Secondly, IOCs that takes on social responsibility and show respect to the environment are likely to survive for long in developing countries. This has become necessary because even though contractual commitments to host governments are honoured and the latter is supposed to take care of the social needs of its people, host governments usually renege on their commitments to the immediate communities. The IOCs may need not take on social responsibility programmes directly but could either set up a separate social enterprise or team up with a known one to implement the social and or environmental programmes.

The last but not the least, is local content. IOCs with programmes to train and employ local manpower are likely to have prolonged stay in developing countries as the latter exert pressure usually backed by civil society to increase local content share of the IOCs' operations.

10. Market options for developing countries

10.1 Policy and regulatory environment

Policy and regulatory environment would be a key driver for the growth of the natural gas market in any country. To ensure growth of the market, the policy and regulatory regime should first of all reflect fair returns to all stakeholders, including government and other stakeholders. The appropriate role of each stakeholder should be spelt out clearly without ambiguity (Energy Commission, 2007).

Secondly, the policy and regulatory regime should be transparent, predictable, clearly defined and open to all investors that meet laid down criteria. The nation's interest is best served by a well defined policy regime that is open to all eligible investors. The policy and regulatory regime should also be simple to administer and not imposing lengthy bureaucracy on private investors. Lengthy bureaucracy could breed corruption since civil and public officials at times take advantage of the long wait-time to promise 'short-cuts' to potential investors. Simplicity in administration is necessary to reduce the costs of compliance to both the investor and relevant government agencies. A policy regime that is loosely defined and subject to discretionary interpretation by public servants can prove costly and can lead to investor uncertainty, the uneven treatment of investment proposals, and can encourage counterproductive behaviour on the part of private sector interests.

10.2 Implementation models

In most developed countries, the Local Distribution Companies (LDCs) are typically owned and operated by municipalities and private enterprises with very little or no public/state sector involvement.

In the developing countries, however they seem to be two models: the **South American Model**, where there is equity participation by both the public and private sectors and the **South-East Asia Model** where most of the transmission and distribution systems are owned and operated by State-owned enterprises. Examples of the South American models are Bolivia (Transredes), Colombia (Promigas), Peru (Suez), Argentina (Metrogas), Chile (Metrogas). Examples of the South East Asia model are found in Pakistan, Bangladesh, India and Thailand.

The South-East Asia models require total funding from the state with all the associated risks. The shortcomings of this model are the inefficiencies in operation and conflicts between the regulatory and operating government entities.

The South American models (and indeed the developed world models), however, require little state involvement.

Since 1990s, most of the countries that subscribed to the South East Asia model are gradually transitioning to the South American Model, one that relies increasingly upon reduced government control and on a more market-responsive pricing climate to encourage foreign and private sector investments. This is expected to push faster the development of the gas sector.

10.3 National transmission and distribution system models

The national transmission and distribution systems in most developing countries have been designed based on (a) the public sector model and (b) the private-public partnership model (Energy Commission, 2007).

10.3.1 Public sector model

A public/state entity with complete operational autonomy, would on behalf of the Government, build and operate the infrastructure. The government would have 100% ownership of the assets. A consortium of LDC and EPCM (Engineering, Procurement, Construction, Management) team may install the facilities and transfer operational control of the network to local employees of the government entity, over a defined period of time under BOT[23] arrangement.

---

[23] Build Operate and Transfer

To be effective, this entity should operate at arm's length from the Government or sector ministry and should be managed by an expert and or, commercially oriented Board of Directors. Despite "best" efforts by most of the developing countries, the public sector models (Figure 4) have mostly been beset with operational failures around the world.

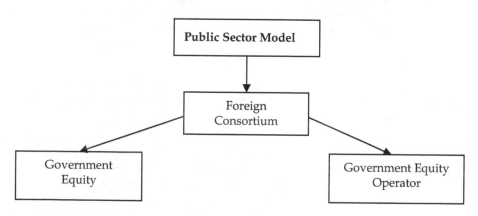

Fig. 4. Public Sector Model.

10.3.2 Public Private Partnership model

The **Public-Private Partnership** model (Figure 5) is usually a joint venture company, with majority private ownership, who builds, own and operate the transmission and distribution network.

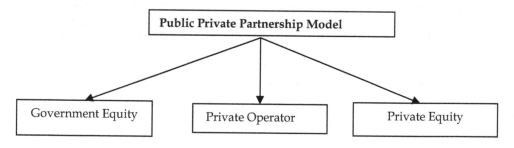

Fig. 5. Public Private Partnership Model.

The government typically may have minority or majority equity interest in the assets, however the minor the better since such usually limit the potential political influence. A consortium of LDC and EPCM team usually install the facilities and transfer operational control of the network to a private local company, over an agreed time period. The private consortium could be a partnership of local and foreign entities.

Table 11 summarizes the main characteristics between state or public ownership and public-private partnership (PPP).

ITEM	STATE/PUBLIC OWNERSHIP	PUBLIC-PRIVATE PARTNERSHIP
Set up	100% Government	Joint venture
Operation	Government entity	Private operating entity
Global Models	Asian Model: e.g. *India, Bangladesh, China, Thailand*	South American Model: e.g. *Bolivia, Colombia, Chile, Peru*
Observed Operational Efficiency	Inefficient, Government interference, subsidies, poor operational oversight	More efficient, commercially driven
System Growth	Expansion of low capacity system likely	Expansion only likely for high capacity load factor systems
Industry Rate of Maturity	Low rate of maturity	High rate of maturity

Table 11. Main characteristics between state/public ownership and public-private partnership.

Most foreign investors would prefer to enter into a joint venture arrangement with a State-owned entity to facilitate certain aspects of infrastructure development, such as land acquisition and landowner issues. A joint venture arrangement with the State could also reduce the risk associated with particular investments.

11. Conclusions

The global energy sector is beset with uncertainties in terms of supply security, development cost, greenhouse gas emissions and other environmental pollution. The flexibility of natural gas as a fuel, its lower carbon dioxide emission compared with other fossil fuels, its emerging global abundant reserves, relatively quick and lower development field cost make natural gas the most favourable fossil fuel in recent times.

The gas industry with its traditional structure of rigid regulations is being replaced by a regime that relies on market forces and is redefining the future gas industry. The structural and regulatory changes that are affecting the global natural gas transmission sector are designed to create competition, expand regional and international gas trade, and to reform the regulation of the transmission and distribution functions to allow for non-discriminatory access to pipelines. Concurrently, these structural changes are being accompanied by ownership changes in light of the global trend towards privatization. Amongst a series of developments, the most evident are the changes in asset structure of the industry, the changing role of gas pipelines, emergence of natural gas commodity markets, fuel switching capability and competition and new pricing mechanism.

In many instances, gas is stranded because of a lack of sufficient demand in locations where the gas is produced. Stranding of gas in locations far removed from big consuming markets calls for construction of pipelines or use of LNG vessels depending upon the distance and location of the consuming market. To complement the global expansion of overland pipeline, LNG investments are also fast growing. For nations that do not have large enough domestic demand relative to the size of their resource base, or that have not developed petrochemicals capacity for conversion of natural gas to other products, LNG is an important means of deriving value for their natural resource endowments through international trade.

Government policies and the type of instruments used to implement the policies however, have impact on gas consumption and fuel choices locally, directly or indirectly. Favourable policies would expand local demand and therefore could add value to the gas being produced by opening up local petrol-chemical industries. With time, there would be significant increase in local demand by the economy and consequently, reduce exports of raw gas. This is the likely market scenario for the future gas industry, particularly, in developing countries with abundant commercial gas resource.

Market and regulatory models for gas markets have been proposed for developing countries.

12. References

BP (2011). Statistical Review of World Energy 2011, www.bp.com

CEE (2006). International Energy Markets, *Economics of the Energy Industries*, Center for Energy Economics, Bureau of Economic Geology of the University of Texas at Austin, (1) pp. 1-19, 2006. www.beg.utexas.edu/energycom

Derman, A. and Johnston, D. (1999). Bonuses Enhance Upstream Fiscal System Analysis, *The Oil and Gas Journal*, Feburary 8, 1999.

Energy Commission (2007). *Natural gas Transmission and Distribution Infrastructure Plan for Ghana*, Energy Commission, Ghana August 2007. www.energycom.gov.gh

Foss, M. (2005a). The Struggle to Achieve Energy Sector Reform in Mexico. Prepared in 2004, for the U.S. Agency for International Development, The Nexus between Energy and Democracy, 2005.

Foss, M. (2005b). Global Natural Gas Issues and Challenges: A Commentary, *The Energy Journal*, Journal of the IAEE, January, 2005.

Johnston, D. (1994). Global petroleum fiscal systems compared by contractor take, *The Oil and Gas Journal*, December 12, pp. 47-50, 1994.

IEA (2011). World Energy Outlook, Special Report – ARE WE ENTERING A GOLDEN AGE OF GAS? International Energy Agency, OECD, 2011, www.iea.org

Mommer, B. (2002). *Global Oil and the Nation State*, Oxford University Press, New York, 2002.

Stevens, P., (2003). *National Oil Companies: Good or Bad? A literature Survey*, National Oil Companies Workshop Presentation, World Bank, Washington, D.C., USA, June 27, 2003.

U.S EIA (2011). Natural Gas: U.S. Natural Gas Pipelines, Energy Information Administration, www.eia.gov

WAGPCo (2011). www.wagpco.com

Major literature consulted in preparation of this chapter

Economics of the Energy Industries, Center for Energy Economics, Bureau of Economic Geology of the University of Texas at Austin, Copyright 2006 Revised Edition 2008. www.beg.utexas.edu/energycom

International Natural Gas Market, Lecture Notes for the International Petroleum Management Program, Institute for Petroleum Development, Austin, Texas 78746, U.SA, 2011 Edition. www.instituteforpetroleumdevelopment.com.

Guide to Natural Gas in Ghana, First Edition, prepared by the Resource Centre for Energy Economics and Regulation, ISSER, University of Ghana, August 2006.

World Energy Outlook, Special Report – ARE WE ENTERING A GOLDEN AGE OF GAS? International Energy Agency, OECD,2011 www.iea.org

Phase Behavior Prediction and Modeling of LNG Systems with EoSs – What is Easy and What is Difficult?

Blanca E. García-Flores[1], Daimler N. Justo-García[2],
Roumiana P. Stateva[3] and Fernando García-Sánchez[1]

[1]*Laboratory of Thermodynamics, Research Program in Molecular Engineering,*
Mexican Petroleum Institute, Mexico, D.F.,
[2]*Department of Chemical and Petroleum Engineering, ESIQIE,*
National Polytechnic Institute, Mexico, D.F.,
[3]*Institute of Chemical Engineering,*
Bulgarian Academy of Sciences, Sofia,
[1,2]*México,*
[3]*Bulgaria*

1. Introduction

In the past decades, natural gas processing has increasingly encompassed low temperatures with the view to recover ethane and heavy components from natural gas. Liquefied natural gas (LNG) is a useful method for storing the gas in a small space at peak-shaving facilities of natural gas distribution companies. The increased density of LNG facilities allows transportation of natural gas via large ocean vessels from gas fields situated far from their potential markets. Also, in the past, natural gas has been an important source of chemical feed stocks like ethane, propane, butane, etc. In order to increase the recovery of these stocks, the gas processes have been shifting to lower temperatures where the recoveries are improved. In all of the above processes and applications of natural gas, knowledge of the LNG systems phase behavior and thermodynamic properties is required for the successful design and operation of LNG plants.

The question of how to predict and model phase equilibria behavior for natural gas systems is far from new and liquid-vapor equilibria problems have been successfully solved for many years now. At present, interest has shifted to systems containing not only species in the simple paraffinic homologous series, but also water, carbon dioxide, hydrogen sulfide, hydrogen, nitrogen– to mention a few.

Not unrelated to this growing variety of species in cryogenic process streams is the occurrence of multiphase phenomena. The situation of interest in natural gas processing is often a methane-rich stream for which there is the question whether a second solvent can alter dramatically the pattern of the phase behavior of the customarily liquid-vapor mixture and cause problems.

The active interest in the use of nitrogen gas to pressurize oil reservoirs to enhance recovery has resulted in natural gas process streams, rich in nitrogen, which are likely to display complex phase behavior. Investigators have studied experimentally ternary prototype LNG systems, containing nitrogen as a second solvent, and a lot of excellent data have been published. However, in order to help further understanding the possible occurrence of multiphase equilibria in LNG process systems, it is necessary to acquire knowledge of their phase behavior and of the variety of critical end point boundaries through an ability to predict, model and calculate them.

There are several aspects of separation/refinement of natural and synthetic gases where multiphase equilibria play a role. The occurrence of liquid-liquid-vapor (LLV) behavior during the recovery of natural gas by low temperature distillation, especially from nitrogen-rich gas mixtures, is just one such example. LLV behavior can also occur in the processing of gases containing high quantities of carbon monoxide that result from coal gasification, as well as in high pressure absorption processes for the removal of either desirable or undesirable components from natural and synthetic gas mixtures. In these latter processes, methanol can be used as the absorber to separate out feed stocks (ethane, ethylene, propane, propylene, etc.), harmful components such as hydrogen sulfide, carbon disulfide, and carbonyl sulfide, or simply unwanted species such as carbon dioxide. Obviously, the formation of a second liquid phase can upset the expected performance of these processes.

In the cryogenic processing of natural gas mixtures, species such as CO_2 and the heavier hydrocarbons can form solids and foul the gas processing equipment. For example, the formation of a solid phase can coat heat exchangers and foul expansion devices, leading to process shutdown and/or costly repairs. Knowledge of the precipitation conditions for gas streams is essential in minimizing downtime for cleanup and repairs. However, the appearance of a solid phase in a process is not always a liability. Off-gas (primary nitrogen) from power plants with light water reactors or from fuel processing plants can contain radioactive isotopes of krypton and xenon, which may be removed by solid precipitation. The phenomenological aspects of LLV/solid-liquid-vapor (SLV) behavior are also interesting because there is a need for a better understanding of the physical nature of the thermodynamic phase space, especially the non-analyticities of critical point region.

The success of the design and operation of separation processes in the oil and gas industry at low temperatures is critically dependent on accurate descriptions of the thermodynamic properties and phase behavior of the concerned multicomponent hydrocarbon mixtures with inorganic gases. Consequently, it is important to apply appropriate models within a thermodynamic modeling framework to predict, describe and validate the complex phase behavior of LNG mixtures. In this case, equations of state (EoS) are usually the primary choice.

The aim in this chapter is to examine and analyze the challenges and difficulties encountered when modeling the complex phase behavior of LNG systems, and to compare the capabilities of two numerical techniques advocated for phase behavior predictions and calculations of complex multicomponent systems. The chapter is organized as follows. In Section 2 the thermodynamics and topography of the phase behavior of LNG systems are presented. Section 3 describes two computational techniques, advocated by us, to predict and model complex phase behavior of nonideal mixtures, and their application to LNG systems. Section 4 is focused to the description of two numerical methods to directly calculate critical end points (K- and L-points). In Section 5 two representatives of the equations of state (EoSs) type thermodynamic models, namely SRK EoS (Soave, 1972) and PC-SAFT EoS (Gross and Sadowski, 2001) are used to represent the equilibrium fluid phases. Results and discussion of the multiphase behavior modeling for selected ternary systems studied are presented in Section 6. Finally, conclusions derived from the work are given in Section 7.

2. Thermodynamics and topography of the phase behavior of LNG systems

Though only a limited number of immiscible binary systems (methane–n-hexane, methane–n-heptane, to name the most prominent ones) are relevant to natural gas processing, LLV behavior can and does occur under certain conditions in ternary and higher LNG systems even when none of the constituent binaries themselves exhibit such behavior. It is also known that the addition of nitrogen to miscible LNG systems can induce immiscibility and this necessarily affects the process design for these systems.

The qualitative classification of natural gas systems and the topography of the multiphase equilibrium behavior of the systems in the thermodynamic phase space and the nature of the phase boundaries will be addressed in this section.

Kohn and Luks (1976, 1977, 1978), who carried out extensive experimental studies on the solubility of hydrocarbons in LNG and NGL (natural gas liquids) mixtures, qualitatively classify natural gas systems (or any system) as one of four types (Kohn and Luks, 1976). These types of binary systems show, for instance, that a first type system, methane-n-octane (Kohn and Bradish, 1964), has a solid-liquid-vapor (S-L-V) locus which starts at the triple point of the solute and terminates at a Q-point (S_1-S_2-L-V) near the triple point of the solvent with an S-V gap in the locus. The S-L-V branches terminate with a K-point where the liquid becomes critical with vapor in the presence of a solid. On the contrary, methane-n-heptane (Kohn, 1961), a second type system, has a Q-point (S-L_1-L_2-V) in the central portion of the SLV locus, at which point one loses the L_1 phase and gains the L_2 phase (solvent lean phase). These systems also have a L_1-L_2-V locus running from the Q-point to a K-point where the liquid L_2 becomes critical with vapor in the presence of L_1. The other two types of systems have no discontinuities between the triple points of the solvents; however, one of these types, methane-n-hexane (Shim and Kohn, 1962), a typical third type system, has a L_1-L_2-V locus which is terminated by critical points, points where the two liquid phases, or a liquid and vapor, become critically identical. A typical system of the fourth type is ethane-n-octane (Kohn et al., 1976). The topographical evolution of multiphase equilibria behavior is discussed in detail by Luks and Kohn (1978). Ternary ($N = 3$) or more complex systems exhibit similar types of phenomena, but the loci and exact points of a binary become $N - 1$ and $N - 2$ dimensional spaces in the more complex systems.

It is known that systems rich in methane can exhibit L_1-L_2-V behavior of the second and third type variety (e.g., methane-n-heptane and methane-n-hexane, respectively). However, investigations into ternary S-L-V and more complex systems revealed that L_1-L_2-V behavior can occur in systems whose binary pairs exhibit no immiscibility (Green et al., 1976; Orozco et al., 1977). Hottovy et al. (1981, 1982) observed that systems of the first type (methane-n-octane) could be modified to behave like a second type system by adding a second solvent (e.g., methane-ethane-n-octane, methane-propane-n-octane, methane-n-butane-n-octane, and methane-carbon dioxide-n-octane). Merrill et al. (1983) reported the phase behavior of the ternary systems methane-n-pentane-n-octane and methane-n-hexane-n-octane, which also exhibit L_1-L_2-V immiscibility. Additionally, these authors studied the systems methane-n-hexane-carbon dioxide, in which the carbon dioxide is added to the pair methane-n-hexane (third type) to induce L_1-L_2-V immiscibility.

In five of these seven ternary systems, methane-n-octane-(ethane, propane, n-butane, n-pentane, and carbon dioxide), the immiscibility region is bounded by loci of type K, type Q and LCST (lower critical solution temperature) points, whereas for the system methane-n-hexane-n-octane the region of L_1-L_2-V immiscibility is bounded, apart from the K, Q, and LCST points, by the L_1-L_2-V locus of the methane-n-hexane binary system. In the case of the system methane-n-hexane-carbon dioxide, the immiscibility region is bounded by a locus of K and LCST points and by the L_1-L_2-V locus of the methane-n-hexane binary system.

It should be pointed out that the onset and evolution of LLV behavior in mixtures is related to the evolution of SLV behavior in those same systems. Thus, in natural gas it is often of interest to predict whether a methane-rich stream with one of the solutes (such as an n-paraffin of carbon number four or higher; or benzene, or carbon dioxide) will form a solid phase. The reason behind that is that the presence of a second solvent can considerably change the solubility of the solid solute, if the solute is a hydrocarbon; this also occurs to a lesser extent if the solute is carbon dioxide.

On the other hand, the presence of nitrogen as a second solvent reduces the solubility of solids in methane, ethane, and mixtures thereof. Furthermore, the addition of nitrogen to miscible LNG systems can induce immiscibility. The number of nitrogen binary systems relevant to LNG that exhibit LLV immiscibility is few – nitrogen-ethane and nitrogen-propane, being among the most prominent ones. However, LLV phenomenon has been observed at certain conditions in many ternary and higher realistic nitrogen-rich LNG systems since LLV behavior can and does occur in multicomponent systems even when for none of the constituent binaries themselves an LLV locus is reported.

Another interesting example is the ternary mixture methane + ethane + n-octane; the species in the constituent binary pairs (methane + ethane) and (ethane + n-octane) are too similar in molecular nature to be LLV immiscible, while the pair methane + n-octane is too dissimilar to be immiscible. If a multicomponent mixture is considered to be solute plus solvent in the pseudo-component sense, it can be readily seen why the ternary mixture has a region of LLV behavior. The methane + ethane form a solvent background in which n-octane is immiscible.

The type of the LLV region displayed by a system depends on whether it contains an immiscible binary or not. For a ternary system with no constituent binary LLV behavior present the three-phase region is a "triangular" surface in the thermodynamic phase space with two degrees of freedom, while its boundaries have one. It is bounded from above by a K-point locus; from below by a LCST locus and at low temperatures by a Q-point locus. The systems methane + n-butane + nitrogen (Merrill et al., 1984a) and methane + n-pentane + nitrogen (Merrill et al., 1984b) belong to this class.

A K-point occurs when a liquid phase and a vapor phase become critical in the presence of a heavier noncritical equilibrium liquid phase, whereas an L-point occurs when two liquid phases becomes critical in the presence of a noncritical equilibrium vapor phase. These points, also called upper critical end points (UCEPs) and lower critical end points (LCEPs), are different depending on their location in the pressure-temperature space. That is, an UCEP is always located at a higher temperature and pressure than the LCEP, and when there exists only one critical end point, so that a K-point is always an UCEP point. On the contrary, an L-point can be an UCEP or an LCEP, depending on the global phase behavior of the system.

The system methane + ethane + n-octane does not exhibit immiscibility in any of its binary pairs. Although immiscibility has been reported in binary systems of methane + n-hexane and methane + n-heptane, solutes such as n-octane and higher normal paraffins crystallize as temperature decreases before any immiscibility occurs. On the other hand, with ethane as solvent, solutes beginning with n-C_{19} and higher paraffins demonstrate LLV behavior. Apparently, the addition of modest amounts of ethane to methane creates a solvent mixture exhibiting immiscibility with n-octane.

3. Modeling of the phase behavior of LNG systems

The success of the design and operation of separation processes in the oil and gas industry at low temperatures is critically dependent on the accurate descriptions of the thermodynamic properties and phase behavior of the concerned multicomponent hydrocarbon mixtures with inorganic gases. Phase-split calculations and phase stability analysis in natural gas systems simulation can take up as much as 50 % of the CPU time. In complicated problems it may take even more. Thus, it is important to develop a reliable thermodynamic modeling framework (TMF) that will be able to predict, describe and validate robustly and efficiently the complex phase behavior of LNG mixtures.

The TMF has three main elements: a library of thermodynamic parameters pertaining to pure-substances and binary interactions, thermodynamic models for mixture properties, and algorithms for solving the equilibrium relations. Reliable pure-component data for the main constituents of LNG systems are available experimentally; an equation of state is usually the primary choice for the thermodynamic model. Thus, the focal point of a TMF for phase behavior calculations of LNG systems is the availability of robust methods for thermodynamic stability analysis, and of reliable efficient and effective flash routines for three phase split calculations.

3.1 Computational technique 1

This first technique uses an efficient computational procedure for solving the isothermal multiphase problem by assuming that the system is initially monophasic. A stability test allows verifying whether the system is stable or not. In the latter case, it provides an estimation of the composition of an additional phase; the number of phases is then increased by one, and equilibrium is achieved by minimizing the Gibbs energy. This approach, advocated as a stagewise procedure (Michelsen, 1982b; Nghiem and Li, 1984), is continued until a stable solution is found.

In this technique, the stability analysis of a homogeneous system of composition z, based on the minimization of the distance separating the Gibbs energy from the tangent plane at z, is considered (Baker et al., 1982; Michelsen, 1982a). In terms of fugacity coefficients, φ_i, this criterion for stability can be written as (Michelsen, 1982a)

$$F(\xi) = 1 + \sum_{i=1}^{N} \xi_i \left[\ln \xi_i + \ln \varphi_i(\xi) - h_i - 1 \right] \geq 0 \quad \forall \xi > 0 \tag{1}$$

where ξ_i are mole numbers with corresponding mole fractions as $y_i = \xi_i / \sum_{j=1}^{N} \xi_j$, and

$$h_i = \ln z_i + \ln \varphi_i(\mathbf{z}) \qquad i = 1, \dots, N \tag{2}$$

Equation 1 requires that the tangent plane, at no point, lies above the Gibbs energy surface and this is achieved when $F(\xi)$ is positive in all its minima. Consequently, a minimum of $F(\xi)$ should be considered in the interior of the permissible region $\sum_{i=1}^{N} y_i = 1$, $\forall \mathbf{y} \geq 0$. To test condition 1 for all trial compositions is not physically possible; it is thus sufficient to test the stability at all stationary points of $F(\xi)$ since this function is not negative at all stationary points. Here, the quasi-Newton BFGS minimization method (Fletcher, 1980) was applied to eq 1 for determining the stability of a given system of composition $\mathbf{z}$ at specified temperature and pressure.

Once instability is detected with the solution at $p-1$ phases, the equilibrium calculation is solved by minimization of the following function

$$\underset{n_i^{(\phi)}}{Min} \Delta g = \sum_{\phi=1}^{p} \sum_{i=1}^{N} n_i^{(\phi)} \ln \left(\frac{x_i^{(\phi)} \varphi_i^{(\phi)} P}{P^\circ} \right) \tag{3}$$

subject to the inequality constraints given by

$$\sum_{\phi=1}^{p-1} n_i^{(\phi)} \leq z_i \qquad i = 1, \dots, N \tag{4}$$

and

$$n_i^{(\phi)} \geq 0 \qquad i = 1, \dots, N \, ; \phi = 1, \dots, p-1 \tag{5}$$

where z_i is the mole fraction of the component i in the system, $n_i^{(\phi)}$ ($i = 1, \dots, N$; $\phi = 1, \dots, p-1$) is the mole number of component i in phase ϕ per mole of feed, $x_i^{(\phi)}$ is the mole fraction of component i in phase ϕ, T is the temperature, P is the pressure, and P° is the pressure at the standard state of 1 atm (101.325 kPa). In eq 3 the variables $n_i^{(p)}$, $x_i^{(p)}$, and $\varphi_i^{(p)}$ are considered functions of $n_i^{(\phi)}$.

Equation 3 is solved using an unconstrained minimization algorithm by keeping the variables $n_i^{(\phi)}$ inside the convex constraint domain given by eqs 4 and 5 during the search for the solution. In this case, a hybrid approach to minimize eq 3 is used starting with the steepest-descent method in conjunction with a robust initialization supplied from the stability test to ensure a certain progress from initializations, and ending with the quasi-Newton BFGS method which ensures the property of strict descent of the Gibbs energy surface. A detailed description of this approach for solving the isothermal multiphase problem can be found elsewhere (Justo-García et al., 2008a).

3.2 Computational technique 2

The second approach to calculate multiphase equilibria used here applies a rigorous thermodynamic stability analysis and a simple and effective method for identifying the phase configuration at equilibrium with the minimum Gibbs energy. The rigorous stability analysis is exercised once and on the initial system only. It is based on the well-known tangent plane criterion (Baker et al., 1982; Michelsen, 1982a) but uses a different objective function (Stateva and Tsvetkov, 1994; Wakeham and Stateva, 2004). The key point is to locate all zeros (y^*) of a function $\Phi(y)$ given as

$$\Phi(y) = \sum_{i=1}^{N} \left[k_{i+1}(y) - k_i(y) \right]^2 \tag{6}$$

where

$$k_i(y) = \ln \varphi_i(y) + \ln y_i - h_i \qquad i = 1,...,N \tag{7}$$

with

$$h_i = \ln z_i + \ln \varphi_i(z) \qquad i = 1,...,N \tag{8}$$

and assuming $k_{N+1}(y) = k_1(y)$. Therefore, from eqs 6–8, it follows that $\min \Phi(y) = 0$ when $k_1(y^*) = k_2(y^*) = ... = k_N(y^*)$. The zeros of $\Phi(y)$ conform to points on the Gibbs energy hypersurface, where the local tangent hyperplane is parallel to that at z. To each zero y^*, a number k^* (equal for each y_i^*, $i = 1,...,N$ of a zero of the function) corresponds, such that

$$k_i^* = \ln y_i^* + \ln \varphi(y^*) - h_i \qquad i = 1,...,N \tag{9}$$

Furthermore, the number k^*, which geometrically is the distance between two such hyperplanes, can be either positive or negative. A positive k^* corresponds to a zero, which represents a more stable state of the system, in comparison to the initial one; a negative k^*, a more unstable one. When all calculated k^* are positive, the initial system is stable; otherwise, it is unstable.

It is widely acknowledged that the task to locate all zeros of the tangent-plane distance function (TPDF), $\Phi(y)$ in this particular case, is extremely challenging because a search over the entire composition space is required. The search is further complicated by the existence of a number of trivial solutions, corresponding with the number of equilibrium phases present (Zhang et al., 2011). The specific form of $\Phi(y)$ (its zeros are its minima) and the fact that it is easily differentiated analytically, allows the application of a non-linear minimisation technique for locating its stationary points, and in their works Stateva and Tsvetkov (1994) and Wakeham and Stateva (2004) used the BFGS method with a line-search (Fletcher, 1980). The implementation of any non-linear minimization technique requires a set of "good" initial estimates, and the BFGS method is no exception. All details of the organization and implementation of the initialization strategy employed by the stability analysis procedure are given elsewhere (Stateva and Tsvetkov 1994) and will not be discussed here.

Thus, as discussed by Wakeham and Stateva (2004), a method has been created which leads, in practice, to an "extensive" search in the multidimensional composition space. It has proved to be extremely reliable in locating *almost all* zeros of $\Phi(\mathbf{y})$ at a reasonable computational cost. The term "almost all" zeros is used because there is no theoretically-based guarantee that the scheme will always find them all. If, however, a zero is missed, the method is self-recovering. Furthermore, the TPDF is minimized *once only*, which is a distinct difference from the approach that stage-wise methods generally adopt. Since stability analysis on its own cannot determine unequivocally which is the stable phase configuration of a system (identified as unstable at the given temperature and pressure), it is suggested to run a sequence of two-phase flash calculations to determine the correct number of the phases at equilibrium, and the distribution of the components among the phases.

4. Calculation of K- and L-points

Ternary systems which exhibit $L_1 - L_2 - V$ behavior but don't exhibit such behavior in their constituent binaries have the immiscibility region bounded by a K ($L_1 - L_2 = V$)-point locus, a LCST ($L_1 = L_2 - V$) locus, and a Q ($S - L_1 - L_2 - V$)-point locus. Ternary systems which have immiscibility in a constituent binary can have boundaries similar to those mentioned above, besides the intrusion of the binary $L_1 - L_2 - V$ locus on the ternary $L_1 - L_2 - V$ region. The K-point and LCST loci can intersect at a tricritical point where the three phases become critical; i.e., $L_1 = L_2 = V$.

Needless to say that it is costly and time-consuming to determine the K and LCST loci in ternary systems experimentally; thus, the availability of appropriate algorithms and numerical routines in the third element of the TMF that will allow the prediction and reliable location in the thermodynamic phase space of such points, is indispensable in the study of the complex phase behavior of LNG model systems. Among the several such algorithms published in the open literature we have chosen to outline briefly and implement those of Gregorowics and de Loos (1996) and Mushrif and Phoenix (2006). In our choice we have been guided by the fact that the above algorithms can successfully predict the K- and L-points of binary and ternary systems. We will thus test their robustness and efficiency in the locating critical end points in LNG model systems.

4.1 Gregorowicz and de Loos' algorithm

In a study on the modeling of the three-phase LLV region for ternary hydrocarbon mixtures with the SRK EoS, Gregorowicz and de Loos (1996) proposed a procedure for finding K- and L-points of ternary systems, based on the solution of thermodynamics conditions for the K- and L-point using the Newton iteration technique and starting points carefully chosen. They applied their procedure to calculate the K- and L-point loci for two ternary systems, namely, $C_2 + C_3 + C_{20}$ and $C_1 + C_2 + C_{20}$, in which the constituent binary $C_2 + C_{20}$ exhibit immiscibility. Consequently, the extension of the three-phase LLV region of these systems is bounded by the binary $L_1 - L_2 - V$ locus of the system $C_2 + C_{20}$ and the ternary K-point and L-point loci. Briefly, the strategy followed by these authors to find the K- and L-points of the two ternary systems was the following: (1) calculation of the critical line, the K-point, the three phase line, and the L-point for the system $C_2 + C_{20}$ using thermodynamic conditions, and (2) calculation of the K- and L-point loci for the ternary systems by using as starting points the obtained coordinates of the K- and L-point calculations for the binary system $C_2 + C_{20}$. In this case, to obtain a K- and L-point for a ternary system, the following set of six nonlinear equations,

$$D = \begin{vmatrix} A_{VV} & A_{Vx_1} & A_{Vx_2} \\ A_{Vx_1} & A_{x_1x_1} & A_{x_1x_2} \\ A_{Vx_2} & A_{x_1x_2} & A_{x_2x_2} \end{vmatrix} = 0 \tag{10}$$

$$D^* = \begin{vmatrix} D_V & D_{x_1} & D_{x_2} \\ A_{Vx_1} & A_{x_1x_1} & A_{x_1x_2} \\ A_{Vx_2} & A_{x_1x_2} & A_{x_2x_2} \end{vmatrix} = 0 \tag{11}$$

$$\mu_i^c - \mu_i^\alpha = 0 \qquad i = 1,2,3 \tag{12}$$

$$P^c - P^\alpha = 0 \tag{13}$$

in seven variables, $T, V^c, V^\alpha, x_1^c, x_2^c, x_1^\alpha,$ and x_2^α have to be solved, where α designates V for the L-point or L_2 for the K-point, D and D^* are the two determinants that must be satisfied at a critical point, and μ_i is the chemical potential of component i.

The critical criteria given by eqs 10 and 11 are based on the Helmholtz energy, which can be expressed as (Baker and Luks, 1980),

$$A = \int_V^\infty \left(P - \frac{nRT}{V} \right) dV - RT \sum_{i=1}^N n_i \ln\left(\frac{V}{n_i RT} \right) + \sum_{i=1}^N n_i (U_i^0 - TS_i^0) \tag{14}$$

where n_i is the mole number of component i and V is the system volume. Derivatives of the Helmholtz energy are denoted by a subscript in eqs 10 and 11 (e.g., A_V, A_{x_i}) indicating the differentiation variable (volume V or mole fraction x_i of component i). Details to obtain the elements of determinants D and D^* are given in Baker and Luks (1980).

It is worth mentioning that Gregorowicz and de Loos (1996) calculated the ternary K- and L-points at chosen values of the temperature, which it is important when the experiments are carried out isothermally.

4.2 Mushrif and Phoenix's algorithm

The second approach to calculate K- and L-points was proposed by Mushrif and Phoenix (2006). This approach utilizes an efficient critical point solver and a standard phase stability test within a nested-loop structure to directly locate K- and L-points. The algorithm consists of two nested inner loops to calculate a critical-point temperature and volume at fixed composition z. An outer loop uses the critical point as a test phase, searches for an incipient phase at a trial composition $\hat{n}$, and updates the critical composition to iteratively decrease the tangent plane distance of the incipient phase to zero. The Newton-Raphson method with numerical derivatives was used in both the inner and outer loops.

This algorithm is similar to that proposed by Gauter et al. (1999) to calculate critical end points (CEPs) for ternary systems of carbon dioxide as the near-critical solvent and two low-volatile solutes (1-pentanol or 1-hexanol + n-tridecane) with the PR (Peng and Robinson, 1976) EoS. These authors used the approach of Heidemann and Khalil (1980) to follow a critical line in steps of the mole fraction of the carbon dioxide, searching for a CEP along this line; i.e., searching for the occurrence of an additional phase in zero amount by using the formulation of Michelsen (1982a).

The principal difference between the algorithm of Mushrif and Phoenix and that of Gauter et al. is that the former directly calculate the K- and L-points without following the critical lines.

In the algorithm proposed by Mushrif and Phoenix (2006), the critical criteria are based on the tangent-plane criteria developed by Michelsen and Heidemann (1988) at specified temperature and pressure. Michelsen and Heidemann (1988) also formulated the critical-point criteria in terms of the tangent plane distance based on the Helmholtz energy at fixed temperature and volume. The condition for the stability of a mixture, with respect to a trial phase $\hat{\mathbf{n}}$, is

$$F = \sum_{i=1}^{N} \hat{n}_i \ln\left(f_i/f_{i_0}\right) - V\left(P - P_0\right)/RT \tag{15}$$

where F is the tangent plane distance from the Gibbs energy surface to the hyperplane tangent of the Gibbs energy surface at the composition $\mathbf{z}$.

In eq 15 the sign of F will determine the stability of the test phase; i.e., if $F > 0$ the test phase is stable, if $F = 0$ the test phase is in equilibrium with some alternate phase, and if $F < 0$ the test phase is unstable. Michelsen (1984) developed the criteria for critical points by expanding the tangent plane distance function in a Taylor series around the test point as

$$F = bs^2 + cs^3 + ds^4 + O(s^5) \tag{16}$$

such that $F(0) = 0$ and $\left(dF/ds\right)_{s=0} = 0$ hold at the test point; s being a parameter that defines the distance in composition space from the test point at $s = 0$. As the sign of the tangent plane distance function F determines the stability of the test phase, it is necessary to find the minimum of this function using scaled mole numbers as $X_i = \left(n_i - z_i\right)/z_i^{1/2}$, where z_i are mole fractions in the test phase and n_i are mole fractions in any alternate phase. At the test point $(\hat{\mathbf{n}} = \mathbf{z})$, $\mathbf{X} = 0$ and $F = 0$. Expressions for the first g_i and second derivatives B_{ij} of F with respect to $\mathbf{X}$ are given in Michelsen (1984).

Function F is minimized by varying $\mathbf{X}$ under the constraint that $\mathbf{u}^T\mathbf{X} = s$, where $\mathbf{u}$ is a vector of unit length. By applying the method of Lagrange multipliers, coefficient b can be expressed as $b = (1/2)\mathbf{u}^T\mathbf{B}\mathbf{u}$, regardless of the choice of vector $\mathbf{u}$. The least possible value of coefficient b is obtained by choosing $\mathbf{u}$ as the eigenvector of $\mathbf{B}$ corresponding to the smallest eigenvalue $\lambda_{\min}$; i.e., $\mathbf{B}\mathbf{u} = \lambda_{\min}\mathbf{u}$.

At trial conditions of temperature and volume, matrix $\mathbf{B}$ is calculated and $(\lambda_{min}, \mathbf{u})$ are determined by inverse iteration (Wilkinson, 1965), then $b = \lambda_{min}/2$. If $b = 0$, the system is at the limit of intrinsic stability. At a critical point coefficients b and c in eq 16 are zero for a given eigenvector $\mathbf{u}$ of $\mathbf{B}$ corresponding to the smallest eigenvalue λ_{min}. For the evaluation of coefficient c, Michelsen (1984) showed that this can be determined efficiently from information already available of u_i and g_i.

The solution procedure to calculate a critical point is as follows: since coefficient b is to be a zero eigenvalue of $\mathbf{B}$, then it must be a singular matrix with a zero determinant; i.e., $\det(\mathbf{B}) = 0$, with a vector $\mathbf{u}$ satisfying $\mathbf{Bu} = 0$ ($\mathbf{u}^T\mathbf{u} = 1$)

The criterion of $b = 0$ is met when the matrix $\mathbf{B}$ is singular and is used to find the critical temperature at a fixed composition and volume. The determinant of matrix $\mathbf{B}$ is calculated through a LU decomposition of $\mathbf{B}$ ($\mathbf{LU} = \mathbf{B}$ where $\mathbf{L}$ is lower triangular and $\mathbf{U}$ is upper triangular); i.e., the determinant of $\mathbf{B}$ is the product of the diagonal elements of the LU decomposed matrix. Once the iteration to find the stability limit has converged, the vector $\mathbf{u}$ is determined by inverse iteration technique.

The implementation of the equation-of-state approach for calculating critical points using this procedure requires that temperature T and volume V are iterated in a nested way. That is, based on an initial guess of V, the temperature is determined in a inner loop until the determinant of the matrix $\mathbf{B}$ becomes equal to zero; then the convergence criterion for the coefficient c is checked. If this coefficient, evaluated at the stability limit, is equal to zero, the calculation ends; otherwise, a new estimate for the volume is generated in an outer loop and the iteration on T is evaluated again. Once T and V have been obtained, the pressure P is evaluated from the equation of state.

After having calculated the critical point, a noncritical equilibrium phase is searched for at constant temperature and pressure conditions. Mushrif and Phoenix (2006) used the stability test implemented by Michelsen (1982a) using the critical composition $\mathbf{z}$ as the reference phase and $\mathbf{Y} = (Y_1,...,Y_N)^T$ as the unnormalized trial phase composition with corresponding mole fractions as $y_i = Y_i / \sum_{j=1}^N Y_j$.

The tangent plane to the Gibbs energy surface at the trial composition is parallel to the reference-phase tangent plane (critical phase) when

$$\ln f_i - \ln f_i^{crit} - \theta = \ln(y_i\varphi_i P) - \ln f_i^{crit} - \theta = 0 \tag{17}$$

where $\theta = -\ln(\sum_{i=1}^N Y_i)$ is the dimensionless distance between the two tangent planes and φ_i is the fugacity coefficient evaluated at composition $\mathbf{y}$.

If $\theta = 0$, the trial phase is in equilibrium with the reference phase; if $\theta > 0$, the trial phase is an incipient phase, and if $\theta < 0$, the reference phase is unstable. By combining eq 17 with the definition of θ, the set of N equations to solve for a stationary point can be written as

$$g_i = Y_i - \exp\left[\ln f_i^{crit} - \ln \varphi_i(\mathbf{y}) - \ln P\right] = 0 \qquad i = 1,...,N \tag{18}$$

which can efficiently be solved by Newton iteration or through a minimization method.

A K- or L-point is found if $\theta = 0$. When θ does not meet the convergence criterion (e.g., $\theta^2 < 10^{-12}$), the critical-phase composition of the lightest component (component 1) is updated using the Newton-Raphson method as

$$z_1^{(k+1)} = z_1^{(k)} - \frac{\theta^{(k)}}{(d\theta/dz_1)^{(k)}} \tag{19}$$

where the derivative $d\theta/dz_1$ is approximated by perturbing the critical composition, recalculating θ from eq 17 and calculating the finite difference analogue of the derivative. Mushrif and Phoenix (2006) have pointed out that failure of the algorithm can occur when a critical composition is updated to a value where no stationary point exist other than the trivial solution.

To calculate a K- or L-point using this algorithm, it is necessary to provide appropriate initial estimates of composition, temperature, and volume. In this case, good initial guesses for critical temperatures and volumes were, depending on the type of calculation, the same as those used by Heidemann and Khalil (1980). The success of the algorithm to locate a K- or L-point strongly depend on (1) the binary interaction parameters used in the equation of state and (2) the initial critical composition $z^{(0)}$. However, it would seem that the value of the initial critical composition significantly affects the successful convergence of the method to locate a K- or L-point.

5. Thermodynamic models

Modeling of the complex phase behavior of LNG systems requires a suitable thermodynamic model and a robust and efficient computational algorithm for performing phase stability and multiphase flash calculations interwoven in the second element of the TMF. Regarding the thermodynamic models, the SRK EoS and the PC-SAFT EoS have received wide acceptance in the industry because of their ability to predict accurately the phase behavior of oil-gas systems.

5.1 The SRK equation of state

The explicit form of the SRK equation of state (Soave, 1972) can be written as

$$P = \frac{RT}{v-b} - \frac{a(T)}{v(v+b)} \tag{20}$$

where constants a and b for pure-components are related to

$$a = 0.42747 \frac{R^2 T_c^2}{P_c} \alpha(T_r) \quad ; \quad b = 0.08664 \frac{RT_c}{P_c} \tag{21}$$

where $\alpha(T_r)$ is expressed in terms of the acentric factor ω as

$$\alpha(T_r) = \left[1 + \left(0.480 + 1.574\omega - 0.176\omega^2\right)\left(1 - T_r^{1/2}\right)\right]^2 \tag{22}$$

For mixtures, constants a and b are given by

$$a = \sum_{i=1}^{N}\sum_{j=1}^{N} x_i x_j a_{ij} \quad ; \quad b = \sum_{i=1}^{N} x_i b_i \tag{23}$$

and a_{ij} is defined as

$$a_{ij} = \left(1 - k_{ij}\right)\sqrt{a_i a_j} \quad ; \quad k_{ij} = k_{ji} , \, k_{ii} = 0 \tag{24}$$

where k_{ij} is an adjustable interaction parameter characterizing the binary formed by components i and j.

Eq 20 can be written in terms of compressibility factor, $Z = Pv/RT$, as

$$Z^3 - Z^2 + \left(A - B - B^2\right)Z - AB = 0 \tag{25}$$

where $A = aP/(RT)^2$ and $B = bP/(RT)$.

The expression for the fugacity coefficient, $\varphi_i = f_i/y_i P$, is given by

$$\ln\varphi_i = \frac{b_i}{b}(Z-1) - \ln(Z-B) - \frac{A}{B}\left(\frac{2\sum_{j=1}^{N} x_j a_{ij}}{a} - \frac{b_i}{b}\right)\ln\left(1 + \frac{B}{Z}\right) \tag{26}$$

5.2 The PC-SAFT equation of state

In the PC-SAFT EoS (Gross and Sadowski, 2001), the molecules are conceived to be chains composed of spherical segments, in which the pair potential for the segment of a chain is given by a modified square-well potential (Chen and Kreglewski, 1977). Non-associating molecules are characterized by three pure component parameters: the temperature-independent segment diameter σ, the depth of the potential ε, and the number of segments per chain m.

The PC-SAFT EoS written in terms of the Helmholtz energy for an N-component mixture of non-associating chains consists of a hard-chain reference contribution and a perturbation contribution to account for the attractive interactions. In terms of reduced quantities, this equation can be expressed as

$$\tilde{a}^{res} = \tilde{a}^{hc} + \tilde{a}^{disp} \tag{27}$$

The hard-chain reference contribution is given by

$$\tilde{a}^{hc} = \bar{m}\,\tilde{a}^{hs} - \sum_{i=1}^{N} x_i(m_i - 1)\ln g_{ii}^{hs}(\sigma_{ii}) \tag{28}$$

where $\bar{m} = \sum_{i=1}^{N} m_i$ is the mean segment number in the mixture

The Helmholtz energy of the hard-sphere fluid is given on a per-segment basis as

$$\tilde{a}^{hs} = \frac{1}{\zeta_0}\left[\frac{3\zeta_1\zeta_2}{(1-\zeta_3)} + \frac{\zeta_2^{3}}{\zeta_3(1-\zeta_3)^2} + \left(\frac{\zeta_2^{3}}{\zeta_3^{2}} - \zeta_0\right)\ln(1-\zeta_3)\right] \tag{29}$$

and the radial distribution function of the hard-sphere fluid is

$$g_{ij}^{hs} = \frac{1}{(1-\zeta_3)} + \left(\frac{d_i d_j}{d_i + d_j}\right)\frac{3\zeta_2}{(1-\zeta_3)^2} + \left(\frac{d_i d_j}{d_i + d_j}\right)^2\frac{2\zeta_2^{2}}{(1-\zeta_3)^3} \tag{30}$$

with ζ_n defined as

$$\zeta_n = \frac{\pi}{6}\rho\sum_{i=1}^{N} x_i m_i d_i^{n} \qquad n = 0,1,2,3 \tag{31}$$

The temperature-dependent segment diameter d_i of component i is given by

$$d_i = \sigma_i\left[1 - 0.12\exp\left(-3\frac{\varepsilon_i}{kT}\right)\right] \tag{32}$$

where k is the Boltzmann constant and T is the absolute temperature.

The dispersion contribution to the Helmholtz energy is given by

$$\tilde{a}^{disp} = -2\pi\rho\, I_1(\eta,\bar{m})\overline{m^2\varepsilon\sigma^3} - \pi\rho\bar{m}\left(1 + Z^{hc} + \rho\frac{\partial Z^{hc}}{\partial\rho}\right)^{-1} I_2(\eta,\bar{m})\overline{m^2\varepsilon^2\sigma^3} \tag{33}$$

where Z^{hc} is the compressibility factor of the hard-chain reference contribution, and

$$\overline{m^2\varepsilon\sigma^3} = \sum_{i=1}^{N}\sum_{j=1}^{N} x_i x_j m_i m_j \left(\frac{\varepsilon_{ij}}{k_BT}\right)\sigma_{ij}^{3} \tag{34}$$

$$\overline{m^2\varepsilon^2\sigma^3} = \sum_{i=1}^{N}\sum_{j=1}^{N} x_i x_j m_i m_j \left(\frac{\varepsilon_{ij}}{k_BT}\right)^2\sigma_{ij}^{3} \tag{35}$$

The parameters for a pair of unlike segments are obtained by using conventional combining rules

$$\sigma_{ij} = \frac{1}{2}\left(\sigma_i + \sigma_j\right) \quad ; \quad \varepsilon_{ij} = \sqrt{\varepsilon_i\varepsilon_j}\left(1 - k_{ij}\right) \tag{36}$$

where k_{ij} is a binary interaction parameter, which is introduced to correct the segment-segment interactions of unlike chains.

The terms $I_1(\eta,\bar{m})$ and $I_2(\eta,\bar{m})$ in eq 33 are calculated by simple power series in density

$$I_1(\eta,\bar{m}) = \sum_{i=0}^{6} a_i(\bar{m})\eta^i \quad ; \quad I_2(\eta,\bar{m}) = \sum_{i=0}^{6} b_i(\bar{m})\eta^i \tag{37}$$

where the coefficients a_i and b_i depend on the chain length as given in Gross and Sadowski (2001).

The density to a given system pressure P^{sys} is determined iteratively by adjusting the reduced density η until $P^{calc} = P^{sys}$. For a converged value of η, the number density of molecules ρ, given in Å⁻³, is calculated from

$$\rho = \frac{6}{\pi}\eta\left(\sum_{i=1}^{N} x_i m_i d_i^3\right)^{-1} \tag{38}$$

Using Avogadro's number and appropriate conversion factors, ρ produces the molar density in different units such as $kmol \cdot m^{-3}$.

The pressure can be calculated in units of $Pa = N \cdot m^{-2}$ by applying the relation

$$P = ZkT\rho\left(10^{10}\frac{Å}{m}\right)^3 \tag{39}$$

from which the compressibility factor Z, can be derived. The expression for the fugacity coefficient is given by

$$\ln\phi_i = \tilde{a}^{res} + \left(\frac{\partial\tilde{a}^{res}}{\partial x_i}\right)_{T,v,x_{j\neq i}} - \sum_{k=1}^{N}\left[x_k\left(\frac{\partial\tilde{a}^{res}}{\partial x_k}\right)_{T,v,x_{j\neq k}}\right] + (Z-1) - \ln Z \tag{40}$$

In eq 40, the partial derivatives with respect to mole fractions are calculated regardless of the summation relation $\sum_{i=1}^{N} x_i = 1$.

6. Results and discussion

Experimental data reported by Llave et al. (1987) for the system nitrogen + methane + ethane, by Hottovy et al. (1981) for the system methane + ethane + n-octane, and by Fall and Luks (1988) for the system carbon dioxide + nitrogen + n-nonadecane, were used to test and compare the robustness, efficiency and reliability of the two computational techniques and the SRK and PC-SAFT EoS thermodynamic models embedded in the respective elements of the TMF. The prediction and modeling of the phase behavior of these systems demonstrates in a clear-cut way the usual numerical difficulties encountered in the process.

The binary interaction parameters used with the PC-SAFT equation were taken from García-Sánchez et al. (2004) and from Justo-García et al. (2008b), while those used with the SRK EoS were taken from Knapp et al. (1982). Some of the interaction parameters were also obtained from the minimization of the sum of squares of the differences between experimental and calculated bubble-point pressures.

The binary interaction parameters employed are: $k_{C_1-C_2} = -0.0078$, $k_{C_1-C_3} = 0.009$, $k_{C_2-C_3} = 0.0170$, $k_{N_2-C_1} = 0.0278$, $k_{N_2-C_2} = 0.0407$, $k_{N_2-C_{19}} = 0.2714$, $k_{CO_2-N_2} = -0.0205$, and $k_{CO_2-C_{19}} = 0.1152$ for the SRK equation, and $k_{C_1-C_2} = -0.0207$, $k_{C_1-C_3} = 0.0168$, $k_{C_2-C_3} = 0.0195$, $k_{N_2-C_1} = 0.0307$, $k_{N_2-C_2} = 0.0458$, $k_{N_2-C_{19}} = 0.1608$, $k_{CO_2-N_2} = 0.0080$, and $k_{CO_2-C_{19}} = 0.1551$ for the PC-SAFT equation, respectively. The components' physical properties required for the calculations performed with the SRK EoS were taken from DIPPR (Rowley et al., 2006) while the three pure-component parameters (i.e., temperature independent segment diameter σ, depth of the potential ε, and number of segments per chain m) of these compounds for the PC-SAFT equation of state were taken from Gross and Sadowski (2001).

6.1 The nitrogen + methane + ethane system

The three-phase VLL region displayed by this ternary system is bounded from above by a K-point locus, from below by a lower critical solution temperature LCST locus, at low temperatures by a Q-point locus, and, due to the fact that this system contains a binary pair (nitrogen + ethane) which exhibits LLV behavior, its LLV space is truncated. In this case, the partially miscible pair nitrogen + ethane spans the LLV space from a position of the LCST locus to a position on the Q-point space. Because methane is of intermediate volatility compared with nitrogen and ethane, it creates a three-phase LLV space which extends from the binary LLV locus upward in temperature. The topographical nature of the regions of immiscibility for the system nitrogen + methane + ethane is shown in Fig. 1. In this figure it can be seen that the L-L=V and L=L-V critical end-point loci intersect at a tricritical point (L=L=V).

Fig. 2 presents the experimental and calculated L_1-L_2-V phase behavior (in terms of L_1-L_2 nitrogen mole fraction data) for the nitrogen + methane + ethane system at 135 K and different pressures. This figure shows a reasonable agreement between the experimental values of liquid phases L_1 and L_2 and those predicted with both models. Notwithstanding, although the LLV calculations performed with both equations up to a position near the K-point (about 41.25 bar with both models), this point is away from the experimental one (43.05 bar at 135 K).

An attempt to directly calculate either the K- or L-point for this ternary system using the algorithm of Mushrif and Phoenix (2006) was carried out. However, the algorithm was not able to give correct values of these critical end points. This is because the algorithm is strongly initialization dependent and hence gives different values of these points, depending on the initial guess of the critical composition, which meets the convergence criterion. Our preliminary results show that the algorithm advocated by Gregorowicz and de Loos (1996) is more stable than that of Mushrif and Phoenix, even if it also depends on the initial values of the critical composition.

Fig. 2 also shows that at pressures away from the LCST point, the PC-SAFT model gives a better representation of the experimental compositions for liquid phase L_1 while both

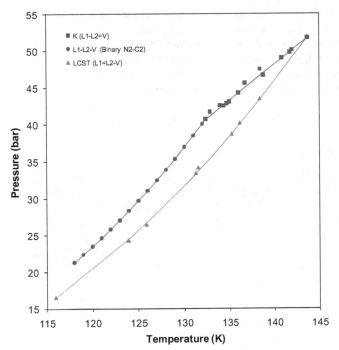

Fig. 1. P-T space of the boundaries of the three-phase L_1-L_2-V regions for the system nitrogen + methane + ethane. Experimental data from Llave et al. (1985, 1987).

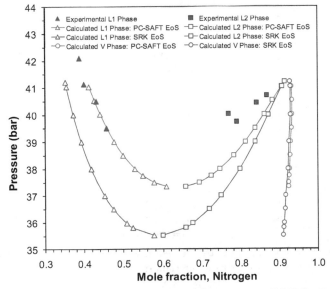

Fig. 2. Comparison of L_1 and L_2 compositional data of nitrogen at 135 K for the system nitrogen + methane + ethane. Experimental data from Llave et al. (1987).

models agree with each other for the liquid L_2 phase but represent the experimental data very closely. Since there are not experimental data below 22.16 bar, comparisons of the models with experiment in the region of the LCST are not possible. However, according to the predictions with both models, the estimated LCST point with the PC-SAFT model (37.20 bar) seems to be closer to the "hypothetical" experimental LCST point (38.37 bar) in comparison with the LCST point estimated from the SRK model (35.45 bar). Of course, these discrepancies can be due to the fact of using binary interaction parameters determined from VL equilibrium data, which, apparently, led to less accurate results.

6.2 The methane + ethane + *n*-octane system

The three-phase LLV region displayed by the methane + ethane + *n*-octane ternary system (a surface in the thermodynamic phase space with two degree of freedom) is bounded from above by a K-point locus (L-L=V), from below by a lower critical solution temperature LCST locus (L=L-V), and at low temperatures by a Q-point locus (S-L-L-V). For the three components in this system, there is no binary immiscibility. The topographical nature of the regions of immiscibility for this system is shown in Fig. 3, where symbols are the experimental data given by Hottovy et al. (1981) identifying the boundaries of the three phase LLV region for this ternary system. This Figure shows also that the L-L=V and L=L-V critical end-point loci intersect at a tricritical point (L=L=V) at the upper temperature limit.

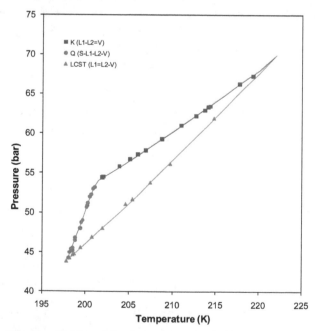

Fig. 3. P-T space of the boundaries of the three-phase L_1-L_2-V regions for the system methane + ethane + *n*-octane. Experimental data from Hottovy et al. (1981).

Following the immiscibility region, a single temperature was chosen to test the capabilities of the PC-SAFT with computational technique 1, and the SRK EoS with computational technique 2, to predict the phase behavior for the system methane + ethane + n-octane. Fig. 4 compares the performance of the two methods at 210 K, and at different pressures on the basis of the experimentally measured and calculated nitrogen mole fractions for the liquid L_1 and L_2 phases by the two thermodynamic models employed. In this case, the predictions of the PC-SAFT model are closer to the experimental composition values than those of the SRK model. However, it should be mentioned that it was not possible to continue the calculations with this model to approach either the K- or L-point because the three-phase LLV triangles become so very narrow as pressure decreases or increases that it is extremely difficult to determine an appropriate global composition able to separate this mixture into three-phase LLV equilibria. On the other hand, because the three-phase LLV triangles predicted with the SRK model are wider than those predicted with the PC-SAFT model, it was easier to get a good initial global composition to calculate the three-phase LLV equilibria from the LCST point (52.80 bar) to the K-point (59.09 bar) applying technique 2.

An inspection of this figure shows that the predictions of both EoSs don't follow the behavior of the liquid phase L_1 as well as the variation of liquid phase L_2 as pressure decreases. Also, it is interesting to note that although there is not a true experimental value of the LCST at the temperature considered, Fig. 4 indicates that the estimated LCST point with the PC-SAFT model (56.35 bar) is closer to the "experimental" one (56.64 bar) than that obtained with the SRK model (52.80 bar). Nonetheless, the "experimental" K-point (60.19 bar) is closer to the one calculated with the SRK model.

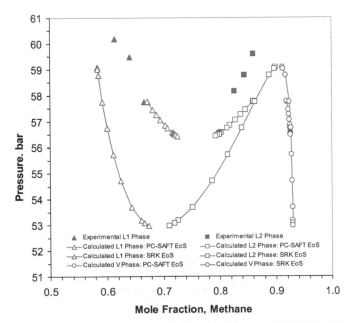

Fig. 4. Comparison of L_1 and L_2 compositional data of nitrogen at 210 K for the system methane + ethane + n-octane. Experimental data from Hottovy et al. (1981).

Fig. 4 also shows that at the different pressures the predicted L_1-L_2-V region (in terms of L_1 and L_2 methane mole fraction data) with the SRK model deviates considerably from the experimental one, while the PC-SAFT model predictions are more reasonable. However, since the interaction parameters for the SRK and PC-SAFT models were determined from binary vapor-liquid equilibrium data, the rather poor fit in this region with either model is not unexpected.

6.3 The carbon dioxide + nitrogen + n-nonadecane system

As mentioned in Section 2, the type of the LLV region displayed by a ternary system depends on whether it contains an immiscible pair or not. In this context, the system carbon dioxide + nitrogen + n-nonadecane exhibits immiscibility in the carbon dioxide + n-nonadecane binary pair (Fall et al., 1985), so that its three-phase region is similar to that exhibited by the system nitrogen + methane + n-pentane (Merrill et al., 1984b). Therefore, the LLV region is "triangular" and is bounded from above by a K-point locus (L-L=V), at low temperatures by a Q-point locus (S-L-L-V), and, from a position of the Q-point locus to a position on the K-point space, by a binary carbon dioxide + n-nonadecane LLV locus.

Fig. 5 presents the experimental pressure-temperature diagram of the LLV space displayed by the system (Fall and Luks, 1986). An examination of the figure shows that this system does not have a LCST (L=L-V) locus and that the Q-point locus terminates at an invariant

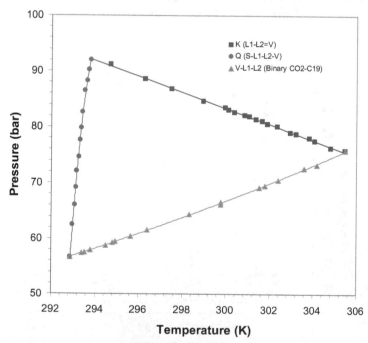

Fig. 5. P-T space of the boundaries of the three-phase L_1-L_2-V regions for the system carbon dioxide + nitrogen + n-nonadecane. Experimental data from Fall et al. (1985) and Fall and Luks (1986).

point of the S-L-L=V type. Thus, due to the fact that the carbon dioxide + nitrogen + n-nonadecane system contains a binary pair (carbon dioxide + n-nonadecane) exhibiting LLV behavior, its "triangular" LLV region is a three-sided space without a tricritical point.

A temperature of 297 K was chosen to study this ternary system and the results obtained are presented in Fig. 6. This figure shows that the PC-SAFT model predicts well the experimental carbon dioxide compositions of the liquid L_2 and vapor phases down to the lowest measured pressure of 62.63 bar (i.e., the binary carbon dioxide + n-nonadecane data) for this isotherm. However, the SRK model is superior to the PC-SAFT model in predicting the three phases in equilibrium for this ternary system. In this case, the calculated L_1, L_2, and V phases are close to the experimental ones. Furthermore, though the SRK EoS overpredicts the "experimental" K-point (87.30 bar) by 4.39 bar the performance of the PC-SAFT EoS in this particular case is inferior as it overpredicts by 14.6 bar the "experimental" point.

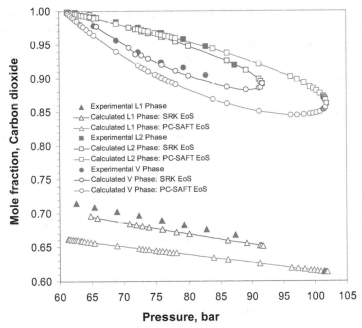

Fig. 6. Comparison of L_1 and L_2 compositional data of carbon dioxide at 297 K for the system carbon dioxide + nitrogen + n-nonadecane. Experimental data from Fall and Luks (1986).

Nevertheless, it should be recalled that all calculations were performed using binary interaction parameters obtained from regression of binary experimental VL data, many of them measured at temperatures higher than that studied here. Furthermore, we are confident that the performance of the corresponding thermodynamic models could be improved considerably provided the interaction parameters were obtained from the regression of the experimental data at three phases. Still, if those sets of interaction parameters are used to predict the phase behavior of a given system at conditions different from the original ones then there is the risk that the equilibria predictions and calculations could either give physical meaningless results or fail altogether.

We also tried to directly calculate the K-points at the temperature considered for this ternary system by using the algorithm of Mushrif and Phoenix (2006); however, once again, the algorithm predicted different values of the critical end points, depending on the initial critical composition. In this case, the strategy to find a critical temperature close to the temperature of study was to use a series of initial compositions. Unfortunately, none of the compositions produced a K-point similar to those obtained from the VLL calculations with both models.

Finally, it should be pointed out that for the sake of comparison both equations of state were interwoven into computational procedure 1 and 2, respectively, and that a series of multiphase flash calculations were carried out at different temperatures and pressures for the three ternary systems studied obtaining the same results for the specific EoS, irrespectively of the computational procedure utilized.

The results obtained showed that there are not any essential differences between, or particular advantages of any of two computational procedures, either in their efficiency, effectiveness, robustness or in their convergence behavior. Thus, it can be said that both procedures 1 and 2 can be used to predict the phase behavior of a wide variety of multicomponent nonideal systems over wide ranges of temperature and pressure.

7. Conclusions

Though there has been much progress and advance in two-phase stability and two-phase split calculations with EoSs, there is still not much progress in three-phase split calculations in natural gas systems despite the large number of publications devoted to the subject. The reason behind that is that the difficulties and challenges are dominating over the easy to perform calculations, if any. Thus, it can be accepted that the algorithms and numerical methods advocated are not robust enough for incorporation in a process simulator. In view of this, the further development of a reliable, robust and efficient TMF and its subsequent approbation on model systems, typical representatives of LNG, is of considerable interest both to scientists and engineers.

On the example of three ternary systems (nitrogen-methane-ethane, methane-ethane-n-octane, and carbon dioxide-nitrogen-n-nonadecane) the capabilities of a TMF, advocated by us, to predict and model complex phase behavior of systems of importance to LNG processing are demonstrated. The TMF employs two numerical techniques which embed different thermodynamic models and stability analysis routines.

The results obtained show that there are many and different challenges and difficulties that are not always possible to overcome completely. For example, the two techniques for multiphase flash calculations cannot always assure steady and non-oscillatory convergence with no tendency towards a strong attraction to the trivial solution, particularly in cases close to the critical lines. Besides, it is known that the phase equilibrium equations are often difficult to converge in the critical region and that the use of inappropriate initial estimates can lead to the trivial solution. In view of this, the availability in a TMF of a robust stability analysis routine that will provide good set of initial estimates for the compositions of possible equilibrium phases and will guarantee steady convergence of the flash routines is of great importance.

Regarding the prediction of K- and L-points with the Gregorowicz and de Loos and Mushrif and Phoenix's methods, it is clear that the both algorithms are strongly initialization dependent. To overcome this problem, for example, Mushrif and Phoenix suggest that the critical phase composition is updated based on the values of parameter θ, calculated from equilibrium phase calculations using the Newton-Raphson iteration. However, during this process of upgrading, the composition may change to a value where there is no phase in equilibrium, particularly when θ differs significantly from iteration to iteration.

The procedure of Gregorowicz and de Loos to calculate K- and L-points seems to be a better fitted method to carry out this task. However, the evaluation of the determinants for solving the conditions of criticality requires the second derivatives of the Helmholtz energy with respect to volume and composition, which makes difficult their evaluation, particularly when these derivatives have to be found analytically applying the PC-SAFT EoS. Still, of course, a possible solution to this problem is to evaluate these derivatives numerically.

Finally, both the PC-SAFT EoS and the RKS cubic EoS are capable of representing with a reasonable accuracy the experimentally observed phase behavior of the ternary systems studied.

8. Acknowledgments

This work was partially supported by the Mexican Petroleum Institute under Research Project D.00406. One of the authors (D. N. J.-G.) gratefully acknowledges the National Polytechnic Institute of Mexico for their financial support through the Project SIP-20110150.

9. References

Baker, L. E. & Luks, K. D. (1980). Critical Point and Saturation Pressure Calculations for Multicomponent Systems. *Society of Petroleum Engineers Journal*, Vol. 20, No. 1, pp. 15–24.

Baker, L. E.; Pierce, A. C. & Luks, K. D. (1982). Gibbs Energy Analysis of Phase Equilibria. *Society of Petroleum Engineers Journal*, Vol. 22, No. 5, pp. 731–742,

Chen, S. & Kreglewski, A. (1977). Applications of the Augmented Van der Waals Theory of Fluids. I. Pure Fluids. *Berichte der Bunsengesellschaft für physikalische Chemie*, Vol. 81, No. 10, pp. 1048–1052.

Fall, J. L. & Luks, K. D. (1986). Effect of Additive Gases on the Liquid-Liquid-Vapor. Immiscibility of the Carbon Dioxide + *n*-Nonadecane Mixture. *Journal of Chemical and Engineering Data*, Vol. 31, No. 3, pp. 332–336.

Fall, D. J.; Fall, J. L. & Luks, K. D. (1985). Liquid-Liquid-Vapor Immiscibility Limits in Carbon Dioxide + *n*-Paraffin Mixtures. *Journal of Chemical and Engineering Data*, Vol. 30, No. 1, pp. 82–88.

Fletcher, R. (1980). *Practical Methods for Optimization. Vol. 1. Unconstrained Optimization*, John Wiley & Sons Ltd, New York.

García-Sánchez, F.; Eliosa-Jiménez, G.; Silva-Oliver, G. & Vázquez-Román, R. (2004). Vapor-Liquid Equilibria of Nitrogen-Hydrocarbon Systems Using the PC-SAFT Equation of State. *Fluid Phase Equilibria*, Vol. 217, No. 2, pp. 241–253.

Gauter, K.; Heidemann, R. A. & Peters, C. J. (1999). Modeling of Fluid Multiphase Equilibria in Ternary Systems of Carbon Dioxide as a Near-Critical Solvent and Two Low Volatile Solutes. *Fluid Phase Equilibria*, Vol. 158-160, pp. 133–141.

Green, K. A.; Tiffin, D. L.; Luks, K. D. & Kohn, J. P. (1976). Solubility of Hydrocarbons in LNG, NGL. *Hydrocarbon Processing*, Vol. 56, No. 5, pp. 251–256.

Gregorowicz, J. & de Loos, Th. W. (1996). Modelling of the Three Phase LLV Region for Ternary Mixtures with the Soave-Redlich-Wong Equation of State. *Fluid Phase Equilibria*, Vol. 118, No. 1, pp. 121–132.

Gross, J. & Sadowski, G. (2001). Perturbed-Chain SAFT: An Equation of State Based on Perturbation Theory for Chain Molecules. *Industrial & Engineering Chemistry Research*, Vol. 40, No. 4, pp. 1244–1260.

Heidemann, R. A. & Khalil, A. M. (1980). The Calculation of Critical Points. *American Institute of Chemical Engineers Journal*, Vol. 26, No. 5, pp. 769–779.

Hottovy, J. D.; Kohn, J. P. & Luks, K. D. (1981). Partial Miscibility of the Methane-Ethane-*n*-Octane. *Journal of Chemical and Engineering Data*, Vol. 26, No. 2, pp. 135–137.

Hottovy, J. D.; Kohn, J. P. & Luks, K. D. (1982). Partial Miscibility Behavior of the Ternary Systems Methane-Propane-*n*-Octane, Methane-*n*-Butane-*n*-Octane, and Methane-Carbon Dioxide-*n*-Octane. *Journal of Chemical and Engineering Data*, Vol. 27, No. 3, pp. 298–302.

Justo-García, D. N.; García-Sánchez, F. & Romero-Martínez, A. (2008a). Isothermal Multiphase Flash Calculations with the PC-SAFT Equation of State. *American Institute of Physics Conference Proceedings*, Vol. 979, pp. 195–214.

Justo-García, D. N.; García-Sánchez, F., Díaz-Ramírez, N. L. & Romero-Martínez, A. (2008b). Calculation of Critical Points for Multicomponent Mixtures Containing Hydrocarbon and Nonhydrocarbon Components with the PC-SAFT Equation of State. *Fluid Phase Equilibria*, Vol. 265, Nos. 1-2, pp. 192–204.

Knapp, H. R.; Döring, R.; Oellrich, L.; Plöcker, U. & Prausnitz, J. M. (1982). *Vapor-Liquid Equilibria for Mixtures of Low Boiling Substances*; DECHEMA Chemistry Data Series, Vol. VI: Frankfurt, Germany.

Kohn, J. P. (1961). Heterogeneous Phase and Volumetric Behavior of the Methane-*n*-Heptane System at Low Temperatures. *American Institute of Chemical Engineers Journal*, Vol. 7, No. 3, pp. 514–518.

Kohn, J. P. & Bradish, W. F. (1964). Multiphase and Volumetric Equilibria of the Methane-*n*-Octane System at Temperatures between -10° and 150°C. *Journal of Chemical and Engineering Data*, Vol. 9, No. 1, pp. 5–8.

Kohn, J. P. & Luks, K. D. (1976). Solubility of Hydrocarbons in Cryogenic LNG and NGL Mixtures. *Research Report RR-22*, Gas Processors Association, Tulsa, Oklahoma.

Kohn, J. P. & Luks, K. D. (1977). Solubility of Hydrocarbons in Cryogenic LNG and NGL Mixtures. *Research Report RR-27*, Gas Processors Association, Tulsa, Oklahoma.

Kohn, J. P. & Luks, K. D. (1978). Solubility of Hydrocarbons in Cryogenic LNG and NGL Mixtures. *Research Report RR-33*, Gas Processors Association, Tulsa, Oklahoma.

Kohn, J. P.; Luks, K. D. & Liu, P. H. (1976). Three-Phase Solid-Liquid-Vapor Equilibria of Binary-*n*-Alkane Systems (Ethane-*n*-Octane, Ethane-*n*-Decane, Ethane-*n*-Dodecane). *Journal of Chemical and Engineering Data*, Vol. 21, No. 3, pp. 360–362.

Llave, F. M.; Luks, K. D. & Kohn, J. P. (1985). Three-Phase Liquid-Liquid-Vapor Equilibria in the Binary Systems Nitrogen+ Ethane and Nitrogen + Propane. *Journal of Chemical and Engineering Data*, Vol. 30, No. 4, pp. 435-438.

Llave, F. M.; Luks, K. D. & Kohn, J. P. (1987). Three-Phase Liquid-Liquid-Vapor Equilibria in the Nitrogen-Methane-Ethane and Nitrogen-Methane-Propane. *Journal of Chemical and Engineering Data*, Vol. 32, No. 1, pp. 14–17.

Luks, K. D. & Kohn, J. P. (1978). The Topography of Multiphase Equilibria Behavior: What Can it Tell the Design Engineer. *Proceedings of the 63rd Annual Convention*, Gas Processors Association, Tulsa, Oklahoma.

Merrill, R. C.; Luks, K. D. & Kohn, J. P. (1983). Three-Phase Liquid-Liquid-Vapor Equilibria in the Methane-*n*-Pentane-*n*-Octane, Methane-*n*-Hexane-*n*-Octane, and Methane-*n*-Hexane-Carbon Dioxide. *Journal of Chemical and Engineering Data*, Vol. 28, No. 2, pp. 210–215.

Merrill, R. C.; Luks, K. D. & Kohn, J. P. (1984a). Three-Phase Liquid-Liquid-Vapor Equilibria in the Methane-*n*-Butane-Nitrogen System. *Advances in Cryogenic Engineering*, Vol. 29, pp. 949–955.

Merrill, R. C.; Luks, K. D. & Kohn, J. P. (1984b). Three-Phase Liquid-Liquid-Vapor Equilibria in the Methane-*n*-Hexane-Nitrogen and Methane-*n*-Pentane-Nitrogen Systems. *Journal of Chemical and Engineering Data*, Vol. 29, No. 3, pp. 272–276.

Michelsen, M. L. (1982a). The Isothermal Flash Problem. Part I. Stability. *Fluid Phase Equilibria*, Vol. 9, No. 1, pp. 1–19.

Michelsen, M. L. (1982b). The Isothermal Flash Problem. Part II. Phase-Split Calculation. *Fluid Phase Equilibria*, Vol. 9, No. 1, pp. 21–40.

Michelsen M. L. (1984). Calculation of Critical Points and Phase Boundaries in the Critical Region. *Fluid Phase Equilibria*, Vol. 16, No. 1, pp. 57–76.

Michelsen, M. L. & Heidemann, R. A. (1988). Calculation of Tri-Critical Points. *Fluid Phase Equilibria*, Vol. 39, No. 1, pp. 53–74.

Mushrif, S. H. & Phoenix, A. V. (2006). An Algorithm to Calculate K- and L-Points. *Industrial & Engineering Chemistry Research*, Vol. 45, No. 26, pp. 9161–9170.

Nghiem, L. X. & Li, Y.-K. (1984). Computation of Multiphase Equilibrium Phenomena with an Equation of State. *Fluid Phase Equilibria*, Vol. 17, No. 1, pp. 77–95.

Orozco, C. E.; Tiffin, D. L.; Luks, K. D. & Kohn, J. P. (1977). Solids Fouling in LNG Systems. *Hydrocarbon Processing*, Vol. 56, No. 11, pp. 325–328.

Peng, D.-Y. & Robinson, D. B. (1976). A New Two-Constants Equation of State. *Industrial & Engineering Chemistry Fundamentals*, Vol. 15, No. 1, pp. 59–64.

Rowley, R. L.; Wilding, W. V.; Oscarson, J. L.; Yang, Y. & Zundel, N. A. (2006). *DIPPR Data Compilation of Pure Chemical Properties Design Institute for Physical Properties*; Brigham Young University, Provo, Utah. http://dippr.byu.edu

Shim, J. & Kohn, J. P. (1962). Multiphase and Volumetric Equilibria of the Methane-*n*-Hexane System at Temperatures between -10° and 150°C. *Journal of Chemical and Engineering Data*, Vol. 7, No. 1, pp. 2–8.

Soave, G. (1972). Equilibrium Constants from a Modified Redlich-Kwong Equation of State. *Chemical Engineering Science*, Vol. 27, No. 6, pp. 1197–1203.

Stateva, R. P. & Tsvetkov S. G. (1994). A Diverse Approach for the Solution of the Isothermal Multiphase Flash Problem. Application to Vapor-Liquid-Liquid Systems. *Canadian Journal of Chemical Engineering*, Vol. 72, No. 4, pp. 722–734.

Wakeham, W. A. & Stateva, R. P. (2004). Numerical Solution of the Isothermal Multiphase Flash Problem. *Review of Chemical Engineering Journal*, Vol. 20, Nos. 1-2, pp. 1–56.

Wilkinson, J. H. (1965). *The Algebraic Eigenvalue Problem*, Clarendon, Oxford.

Zhang, H.; Bonilla-Petriciolet, A. & Rangaiah, G. P. (2011). A Review on Global Optimization Methods for Phase Equilibrium Modeling and Calculations. *The Open Thermodynamics Journal*, Vol. 5, pp. 71-92.

3

The Gas Transportation in a Pipeline Network

Jolanta Szoplik

West Pomeranian University of Technology, Szczecin,
Poland

1. Introduction

During the last few years we have been able to observe increasing development of different types of networks. Energetic, telecommunication, water or gas networks are only a few examples of systems whose main purpose is media transportation from the source (producer, manufacturer) to the target place of use. One of the basic features of network systems is their uniqueness with regard to the structure as well as transmission capability. In practice there are no two identical networks and their uniqueness leads to the necessity of use of individual approach to networks during designing and exploitation stage. Moreover networks have complex structure and any change or modification of their structure while in use is not an easy task. For these reasons flow improvement in current network and optimal exploitation of its capabilities is an important and actual issue.

In case of many real-time systems, examination of their performance in conditions different from currently existing is impossible. In such a situation a mathematical model of such system is constructed which is a kind of simplification of the reality (Kralik et al., 1988; Osiadacz, 1987 and Osiadacz, 2001). However, the model of the system can be used to simulate its behaviour in reaction on extortion. Simulation is an example of an experiment, performed with the help of an appropriate algorithm and a computer. An enormous advantage of computer simulation is its repeatability, and each simulation task in the same initial conditions can be repeated infinitely many times. In real conditions such an experiment is almost impossible to perform, as from technical point of view, it is extremely difficult to perform direct measurement of some values characterizing work of the system. However, conducting a series of simulations for different initial conditions allows to receive various solutions and to choose the best one out of those received. Such an example of using simulation results can be treated as quasi-optimisation which can be accepted as sufficient when the relations between elements of the system and extortion are not well known. The approach which uses mathematical model is a huge simplification of the real system.

A gas network can be an example of a system, where from the technical point of view, it is difficult to perform direct measurements of parameter values characterising flow in pipeline networks. Currently we observe evident increase in gas consumption by the municipal receivers as well as the industry, and this is the reason why gas transportation network from the place of extraction directly to the recipient results in pipelines system becoming more and more complex. Gas in the network is transported under appropriate pressure. Depending on its level we can distinguish three different types of networks: high, middle or low pressure networks (Osiadacz, 1987; Kralik et al., 1988). The analysis of network flow simulation results enables us to estimate such network bandwidth reserves, define possibilities of network expansion directions and possibility of connecting new recipients, calculate maximal value of the gas flow rate in the pipeline of the network or gas parameters (e.g. pressure) in output points. Overpressure layout, velocity of the gas and gas flow rate in each pipeline of the given gas network can be defined on the basis of network gas flow simulation.

In the literature, there are many examples of application various mathematical methods and IT tools, to facilitate solving complex simulation tasks, optimisation or steering in regard to gas flow in the gas pipeline or gas pipelines network.

The subject of several scientific papers was the analysis of flow issues and leakage detection in network systems. The authors stated, that these problems can play an important role in the management of pipeline system and to cause reduce the loss of leakage. Gonzalez et al. (2009) focused on modeling and simulation of gas pipeline network and presented two models derived from the set of partial differential equations and two numerical schemes for integration of such models. Fukushima et al. (2000) proposed and successfully implemented the leak detection system based on a dynamic simulation with wave equation using real operational data on their one of the longest gas pipeline in Japan. Brkić (2009) for construction and calculation of looped gas distribution pipeline network of composite structure with known node gas consumption proposed to use an improvement of Hardy Cross method procedure. Liu et al. (2005) presented an adaptive particle filter to tackle the leak detection and location in gas pipelines. Reddy et al. (2006) used dynamic simulation models (transfer function model) of gas pipeline network for on-line leak detection and identification and they stated, that proposed method was 25 times faster than the explicit finite – difference approach.

The problems of heat exchange between gas flowing through the pipe and the ground analysed Osiadacz & Chaczykowski (2001), Chaczykowski (2010), Ke &Ti (2000), Tao & Ti (1988). Ke & Ti, 2000 and Tao & Ti, 1988 proposed for transient analysis of isothermal gas flow in the pipeline network a new mathematical model based on electrical analogy. These papers treated mainly of gas flow in high pressure network that have much simpler structure than low pressure networks.

However, applying in the research Artificial Neuronal Network (ANN), does not require earlier knowledge of exact relations between particular values, and it is only sufficient to know input and output data. Examples of use of Artificial Neuronal Networks for modelling, steering or optimisation of fluid flow or transport supporting appliances can be found in literature. Zahedi et al. (2009) propose applying the ANN method to predict hydrates forming temperature during gas transportation with gas pipeline. According to the authors results received using the ANN are more adequate than analogous results received with traditional methods. Carvalho et al. (2006) applied the ANN to detect and then to classify faults on side of the gas pipeline to appropriate group. The authors proved that precision of the ANN to detect faults on gas pipeline side is approximately 94%, whereas correctness of recognition of fault type (corrosion or other) is approximately 92%. The ability of the ANN to recognise type of corrosion (internal, external) is clearly smaller (approximately 72%). Silva at al. (2007) in their work introduced proposal of applying the

ANN for analysis of influence of pipeline corrosion defects on chosen parameters characterising the flow. The authors however noticed that there is necessity of verification of work effects with real data. Based on the results from magnetic flux leakage signals, Hwang et al. (2000) presented a new approach for training, hierarchical wavelet basis function neural network for the three-dimensional characterization of defects on the pipeline. Nguyen et al. (2006, 2008) use the Neural Network for forecasting the demand of the hourly gas stream for customer. Experimental data obtain by ANN are used in the Genetic Algorithm to search the optimal combination of compressor scheduling.

The purpose of this chapter is to perform the analysis of results of the steady-state simulation calculation for gas flow in the low pressure gas pipelines network, based on which, two methods of steering the gas stream pressure entering the network will be described. Under discussion will be variability of gas consumption from the network by different recipients groups in seasonal and daily cycle. Air temperature will be important parameter with significant influence on gas consumption by consumers. There will be an algorithm developed to steer gas pressure in form of dependence of gas pressure on stream volume feeding the network, as well as the network characteristic with marked network nodes, where the pressure is lowest. It was proved, that steering gas pressure feeding the network, due to keeping lower gas pressure in the network, can significantly lower the network exploitation costs.

In the present chapter fragment of real low pressure network in Szczecin city (Poland) was used and calculations were performed for real data presenting hourly gas streams leaving 108 nodes of the network in successive hours of four chosen days with various air temperature.

2. The elements of gas pipeline network system

Gas network consists of connected and cooperating with each other objects that transport and distribute natural gas. Fig. 1 presents main objects that compose to the gas network and their main classification. Gas pipelines with equipments are used to transfer and distribute gas fuel to consumers. Polish classification of gas pipeline networks and short characteristics (gas overpressure (p), gas speed (w) and type of pipes material) are presented in Fig. 2. Taking into account the gas stream overpressure, there are four types of gas pipelines. Another classification is, when as classification criteria, the type of pipes material in the network. In this case there are gas pipelines made of polyethylene (PE) or steel. Fig. 3 presents all previously mentioned gas network objects.

Gas station is a set of appliances in the gas network fulfilling separate or simultaneous functions of pressure reduction, stream measurement and gas characteristics or stream division. The gas compressor station raises the gas stream pressure in order to overcome the pressure lost resulting from the frictional gas flow in the pipeline of network.

Turbines in compressor stations can be fed with gas from the gas pipeline or electrically. Gas storehouses are natural containers constructed in rock mass, underground mining excavations or salt caverns, and used to storage gas in periods of lower demand (months of high air temperature) and additional network feed during periods of higher gas demand (months of low air temperature).

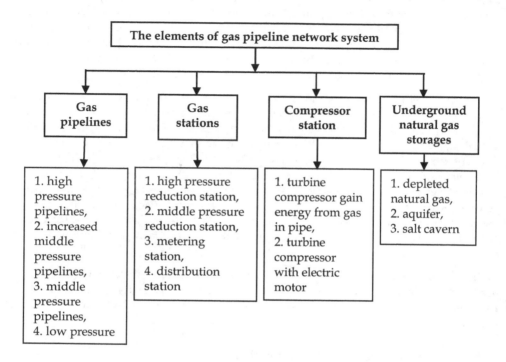

Fig. 1. Technical units composed the gas pipeline network.

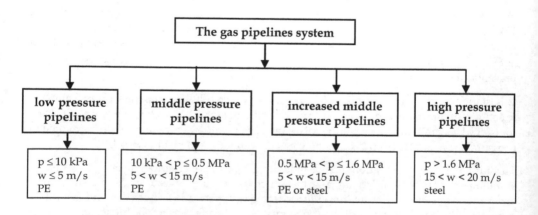

Fig. 2. The main types of gas pipeline networks.

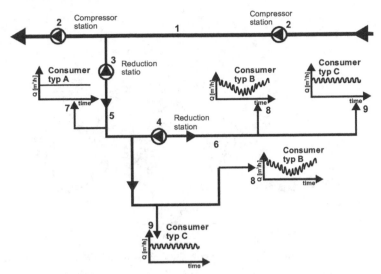

Fig. 3. Fragment of gas network with marked objects and gas use characteristics by different groups of consumer.

Fig. 3 shown that, large amount of natural gas is transported on long distances through the high pressure pipelines network (1). Compressor stations (2) are located on gas pipelines in distance of 100-150 km, that increase initial gas pressure lost during the flow. Whereas natural gas delivered to consumers is characterised with clearly lower overpressure that is received after two stage pressure reduction in reduction stations. High pressure gas reduction station (3) lowers gas pressure from high to medium level. However, in middle pressure gas reduction station (4) is second stage of gas pressure reduction to low level. Gas pipelines network between gas reduction stations (3) and (4) is called (middle pressure gas network) (5), whereas from the gas reduction station (4) distribution of gas with low pressure gas network (6) begins directly to municipal consumers (7), (8), (9).

3. The variety of the gas demand in the year

Natural gas transported with network presented in Fig. 3, is delivered to industrial (7) and municipal (8) and (9) consumers. In the industry the natural gas is used as raw material and main source of methane, whereas in households gas is used to prepare meals and heat water and accommodation. In Fig. 3, three types of gas consumers are marked and presented their gas usage characteristic within a year. The A type (7) is an industrial consumer, who consumes from the network the same gas amount regardless of time of the year and time of the day. The B type (8) and C type (9) are municipal consumers that use gas in households. The B type consumer is characterised with variable gas consumption within daily and seasonal cycle, which means, that gas is mainly used to house heating. The C type consumer uses gas only to prepare meals or possibly to heat water, therefore this kind of variation can be accepted as only daily variation.

Gas consumption by the B type consumers mainly depends on weather and calendar factors. Fig. 4 presents exemplar diagram of gas consumption variation by large group of B and C

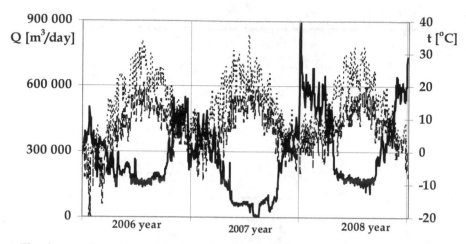

Fig. 4. The change of gas stream Q in days of years 2006, 2007 and 2008; (…) - t_{max} the day temperature, (– –) t_{min} the night temperature, (—) Q daily volumetric gas flow; (Szoplik J., 2010a).

type consumers together, for one of Polish cities in successive days of years 2006, 2007 and 2008 depending on air temperature. The increase of air temperature during summer months (July, August) influences on visible gas consumption decrease and vice versa, low air temperatures (January, February, November and December) cause, that large gas amounts leave the network and the gas is used mainly to heating. Variation of gas consumption from the network results with various network loads.

4. Mathematical model for gas network

The computer programs used for simulation of gas pipeline distribution transport system are based on the gas network mathematical model. The detailed form of the mathematical model for such a system depends on the assumptions of flow as well as the conditions of the network operation. There is no universal formula to describe the flow of gas in a pipeline. The different equations are used depending on the working pressure of the network and the assumptions made with regard to the conditions of the network operation (Osiadacz, 1987; Osiadacz, 2001; Kralik et al., 1988). Three equations describing the gas flow through a pipeline network are derived from the equation of continuity, the equation of motion and the equation of energy (Osiadacz & Chaczykowski, 2001; Ke & Ti, 2000):

$$\frac{\partial \rho}{\partial t} + \frac{\partial (\rho w)}{\partial x} = 0 \tag{1}$$

$$\frac{\partial (\rho w)}{\partial t} + \frac{\partial (\rho w^2)}{\partial x} + \frac{\partial p}{\partial x} + 2\rho w^2 \left(\frac{\lambda}{D}\right) + g\rho \sin \alpha = 0 \tag{2}$$

$$\frac{\partial}{\partial t}\left[(\rho A dx)\left(c_v T + \frac{w^2}{2} + gz\right)\right] + \frac{\partial}{\partial x}\left[(\rho w A dx)\left(c_v T + \frac{p}{\rho} + \frac{w^2}{2} + gz\right)\right] - q\rho A dx = 0 \tag{3}$$

where: ρ is the density of gas, w is the gas flow velocity, λ is the Finning friction coefficient, D is inner diameter of the pipe, α is the angle between the horizon and the direction x, A is the cross-section area of the pipe, c_v is the specific heat at constant volume, q is the specific heat related per unit mass and T is the temperature of gas.

Stationary gas flow in a pipeline does not vary in time and is a special case of dynamic behaviour of gas flow in a pipeline. In this case, the variables in the equations of continuity, motion and the energy are only a function of coordinates and this system is described by the set of nonlinear algebraic equations (Osiadacz, 1987 and Osiadacz, 2001). For the low-pressure gas pipeline network, when the change of the gas pressure and the dynamics of the flow can be neglected, it is suitable to use the steady-state simulation. The more complex the pipeline network is, the more complex the mathematical model of such system is therefore in order to solve this problem in most cases computers are used. In case of gas networks the structure of the pipeline network can be presented by means of the graph theory, which allows simple representation of the structure in terms of the properties of its elements incidence.

A graph (Kralik et al, 1988; Osiadacz, 1987 and Osiadacz, 2001) consists of a set of nodes and a set of branches. The nodes are presented by points whereas the branches by line segments connecting two points. When a branch has a pair of nodes in a certain order the graph is called direct graph. Sometimes the node is connected to itself and this closed path of nodes and branches is called a loop of the graph. In case of the gas pipeline network the boundary nodes are the points where the elements are connected to the whole network and this is the node that belongs to a single element. The second type of node is an internal node of the network which is common to exactly two elements (simple node) or to at least three elements (crossing node). Gas enters the network at node called supplier node and leaves the network at boundary nodes. A branch in gas pipeline network is a pipe with constant diameter and roughness. Taking into account the position of a branch in the network there are boundary and internal branches.

The interconnection of the pipe in the network can also be presented by the branch-nodal incidence matrix $A = [a_{i,j}]_{n \times m}$. The number of rows ($n$) is equal to the number of nodes, but the number of columns (m) is equal to the number of branches (Osiadacz, 1987; Osiadacz, 2001 and Kralik et al., 1988). Each element $a_{i,j}$ of the matrix A is equal to 0 or (+1) or (-1).

$$a_{ij} = \begin{cases} +1- \text{ when branch } (j) \text{ enters node } (i), \\ -1- \text{ when branch } (j) \text{ leaves node } (i), \\ 0- \text{ when branch } (j) \text{ is not connected to node } (i) \end{cases}$$

The incidence of the loops and branches describes the matrix of branch-loop $B = [b_{i,j}]_{k \times m}$. The rows ($k$) in matrix B correspond to the loops while the columns (m) correspond to the branches in the network. The elements $b_{i,j}$ are defined as:

$$b_{ij} = \begin{cases} +1- \text{ when branch } (j) \text{ has the same direction as loop } (i), \\ -1- \text{ when branch } (j) \text{ has opposite direction to loop } (i), \\ 0- \text{ when branch } (j) \text{ is not in loop } (i) \end{cases}$$

Osiadacz (1987); Osiadacz (2001) and Kralik et al. (1988) propose that for the simulation of the steady-state pipeline networks the analogy between fluid and electrical network can be successfully applied. The aim of the simulation of the gas flow is to estimate the values of the flow rates in each pipe of the network and the pressure at each node of the network. The calculated value of flow and pressure must satisfy the flow equation and together with the value of the loads and off-takes gas flow must satisfy the first and second Kirchhoff's laws.

The matrix form for the first and second Kirchhoff's law (Osiadacz, 1987; Osiadacz, 2001 and Kralik et al., 1988) represent the following equations:

$$A_1 \cdot Q = q \quad \text{(I Kirchhoff's law)} \tag{4}$$

$$B \cdot \Delta p = 0 \quad \text{(II Kirchhoff's law)} \tag{5}$$

where:

$A_1 = [a_{ij}]_{(n-n1) \times m}$	– reduced nodal-branch incidence matrix,
$B = [b_{ij}]_{k \times m}$	– loop-branch incidence matrix,
$Q^T = [Q_1, Q_2, \ldots Q_m]$	– vector of flows in the branches,
$q^T = [q_1, q_2, \ldots q_{(n-n1)}]$	– vector of stream at the output nodes,
$\Delta p^T = [\Delta p_1, \Delta p_2, \ldots \Delta p_m]$	– vector of pressure drops in the branches,

Equations (4) and (5) complete with one of the following forms of flow equation (equation 6 or 7) are the matrix form of the mathematical model and describe the gas flow in the network.

$$Q = \Psi(\Delta p) \tag{6}$$

$$\Delta p = \Phi(Q) \tag{7}$$

where: $\psi(\Delta p)$ is the vector of pressure drop functions in the branches and $\Phi(Q)$ is the vector of flow functions

There are two methods which are most frequently used to carry out the simulation of the gas flow in the network. In the nodal method, the equations (4, 5 and 6) are resolved. Initial approximations are made to the nodal pressure and are corrected in the next iterations until the final solution is reached. The loop method is the other way to resolve the mathematical model for the gas flow. Initial approximations are made to the branch flow and next iterations are corrected until the final solution is reached. The loop method provides the solution of the equations (4, 5 and 7).

5. The subject of the analysis in the study

The subject of the analysis was the real low pressure pipeline network which consisted of 319 pipelines of various diameters (from 0.050 m to 0.25 m). The whole length of the pipeline in this network was equal to 4151 m and the overall amount of gas accumulated in the network was equal to 51 m³. The low pressure gas pipeline network operated with overpressure in the range of 1.7 kPa to 2.5 kPa. The operating temperature was 283 K, the relative density of the gas was equal to 0.6. The velocity of the gas was always lower than 5 m/s in each pipe of the network.

The graphic representation of the network analysed in the study consisted of 316 nodes and 319 branches (Szoplik J., 2010b, 2010c). Taking into account the position of nodes in the network is possible to distinguish two types of nodes: boundary and internal nodes. There was 1 supplier node (Fig. 5, point Z1 – gas reduction station), where gas entered the network, 108 nodes, where gas left network (boundary or output nodes) and 207 internal nodes. There were also one input, 108 output and 210 internal branches in the graph of the network.

Boundary branches containing input or output nodes are called input and output branches respectively, but the elements which are neither input nor output elements of the network are called internal elements. The number of loops in this pipeline network was 3. Fig. 5 is the graphic representation of the network used in the study. The detailed data for the graph, diameter and length of each type of the pipe in the network presented in Fig. 5 are collected, respectively, in Table 1 and Table 2. There are only 13 nodes distinctly marked at the graph presented in Fig. 5 (A2; A51; A61, A64, A65, A70, A71, A75, A92, A90, A80, A146, A147). The flow rate at these nodes is divided or the diameter of the pipe is varied.

No	Description	Number
1	Total number of nodes	316
2	- supplier node	1
3	- boundary (output) nodes	108
4	- internal nodes	207
5	Total number of branches	319
6	- output branches	108
7	- internal branches	210
8	Number of loops	3

Table 1. The characteristics of the nodes and branches for the graph presented in Fig. 5.

The mathematical model of gas flow in the pipeline network consisting of equations (4), (5) and (7) was performed with computer program GasNet, used to steady-state simulation of gas flow by means of loop method. Calculations were performed for real fragment of the low pressure gas pipeline network in the city Szczecin and were based on real data characterising hourly values of gas streams which left the network at each of 108 output nodes for different values of gas stream overpressure entering the network in gas reduction station Z1 (Fig. 5). The drop pressure in pipelines constituting network was calculated as a difference of absolute pressure in two adjacent nodes. The friction factor λ was determined according to the guidelines concluded in the norm PN-76/M-34034.

No	D_{nom} [mm]	D_{in} [mm]	s [mm]	L [m]
1	250	204.6	22.7	18.7
2	225	184.0	20.5	687.6
3	180	147.2	16.4	1280.7
4	160	130.8	14.6	80.2
5	125	102.2	11.4	517.1
6	90	73.6	8.2	813.5
7	63	51.4	5.8	735.8
8	50	40.8	4.6	17.8

Table 2. The diameter D, length L and thickness s of the wall for the pipes of the network presented in Fig. 5.

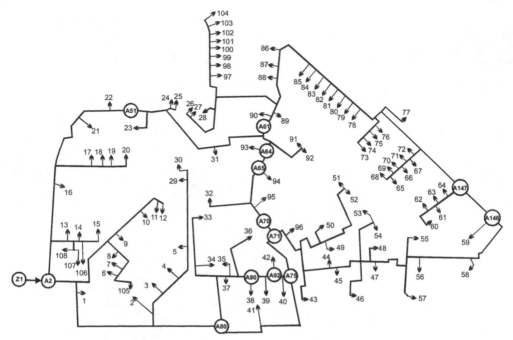

Fig. 5. The graph of the gas pipeline network; Z1 – middle pressure gas reduction station; (Szoplik J., 2010b).

In the case of the low pressure gas pipeline network the drop pressure in each branch was calculated as the difference of the pressure at two adjacent nodes. The friction factor λ in the laminar region ($Re \leq 2300$) is defined by the Hagen-Poliseuille:

$$\lambda = \frac{64\eta}{wD_{in}\rho} = \frac{64}{Re} \tag{8}$$

For the turbulent gas flow the friction factor depends on the Reynolds number as well as the relative roughness of the pipe wall ($e = k/D_{in}$). The friction factor in transitional region ($Re > 4000$ for $e \leq e_{bound}$), depends only on Reynolds number and is described by the implicit relationship Prandtl-Karman:

$$\lambda = \left[2\log \frac{\sqrt{\lambda}\,Re}{2.51} \right]^{-2} \tag{9}$$

Calebrook-White equation is used to calculate the friction factor for the flow in fully turbulent region ($Re > 4000$ and $e > e_{bound}$) when factor λ depends also on the Reynolds number and relative roughness e:

$$\lambda = \left[-2\log \left(\frac{2.51}{Re\sqrt{Re}} + \frac{e}{3.72} \right) \right]^{-2} \tag{10}$$

The boundary relative roughness e_{bound} describes one of the relationships below:

- Filonienko-Altsul

$$e_{bound} = \frac{18 \log Re - 16.4}{Re} \tag{11}$$

- Blasius

$$e_{bound} = 17.85 Re^{-0.875} \tag{12}$$

- Altsul-Ljacer

$$e_{bound} = \frac{23}{Re} \tag{13}$$

5.1 The characteristic of input data

Steady-state gas flow simulation calculations are conducted based on properly prepared input data in form of 108 values of hourly gas streams consumed from the network by consumers in consumption nodes.

Amount of daily gas stream leaving particular node of the network depends inter alia on air average temperature. Based on archive data covering years 2006-2008, linear relations of daily gas consumption in function of air average temperature were developed. Details of defining linear models of gas consumption value from the temperature are described in literature (Szoplik J., 2010a). General form of linear model to define daily gas stream Q_d [m³/day] consumed by consumers from particular node is presented by equation

$$Q_d = a(18-t) + b \tag{14}$$

where: a, b – model constant, individually determined for each node of the network based on real data of gas consumption by consumers, t - average air temperature [°C].

Fig. 6. presents results describing gas stream volume variation Q_d received within twenty four hours in four selected nodes of the network from the Fig. 5 depending on average air temperature t = +18, +8, -4, -16 °C (set according to equation (14)). Comparing results presented in Fig. 6 one can see, that the increase of daily gas stream Q_d leaving the network, caused by temperature increase, is not equal in all nodes of the network, as the number of gas consumers and gas receivers installed at consumers, consuming gas from a particular node of the network is different.

However amount of hourly gas stream Q [m³/h] leaving particular node of the network is defined based on characteristic of percentage gas consumption in particular hours of the day. Such characteristics were developed based on real data describing gas stream flow through reduction and measurement station in successive hours of the day in various days of the year. Fig. 7 presents gas stream size change during different hours of the day with average temperature of -4 °C in four selected nodes of the network from the Fig. 5. However Fig. 8 presents the results in form of hourly stream size Q of the gas leaving the network in node 55 during various hours of four exemplar days, differ with air temperature.

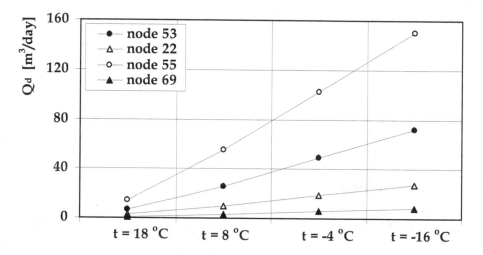

Fig. 6. The effect of air temperature t on the volumetric daily gas stream size Q_d in a given node of the network.

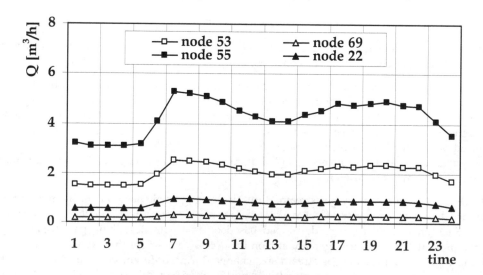

Fig. 7. Change of the gas stream size Q leaving the network in four exemplar nodes of the network in particular hours of the day; air temperature t = -4 °C.

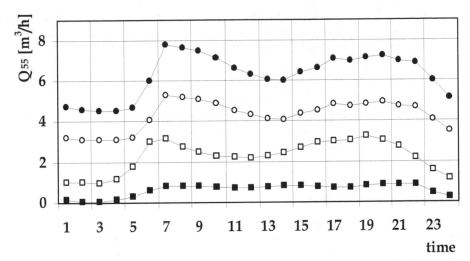

Fig. 8. Change of the gas stream Q_{55} leaving the network in node 55 in particular hours of the day; various values of air temperature; (●) t = -16 °C, (○) t = -4 °C, (□) t = 8 °C, (■) t ≥ 18 °C.

Analysing results presented in Fig. 7 and Fig. 8 one can notice, that size of the hourly gas stream Q depend on air temperature and hour of the day and type of the node in the network. It can also be noticed, that in the same node of the network (i.e. node 55) maximal gas stream size Q_{max} noted on the day with average temperature of -16 °C is nine times larger than that the day with temperature of 18 °C. There are also clear differences between maximal and minimal stream values Q during a day (for the same node). In percentage, larger diversity is noticed during summer days than during winter days. During winter time (t = -16 °C) maximal gas stream size Q_{max} is larger than minimal stream Q_{min} by approximately 70%, however it is two times larger for summer time (t ≥ 18 °C).

6. The results and discussion

Change of the gas stream leaving the network in 108 nodes during season and twenty four hour cycle is reflected on appropriate diversity of the network load. With the GasNet 3.8.1 programme, a steady-state simulation of gas flow in network, for each set of input data (108 values of gas streams) was conducted. Details of mathematical model of gas flow in the network and algorithm, according to which calculations were performed in the GasNet programme, were described in papers (Szoplik J., 2010b, 2010c). For each of four, exemplar days different with average temperature conducted 24 simulations. In total, performed 96 simulation calculations, are presented in this chapter.

In practice, it is accepted, that the network works correctly, when each collection node gas stream of appropriate size is delivered with pressure higher than p ≥ 1700 Pa. Fulfilling above two conditions is possible due to correct work of the reduction gas station (point Z1, Fig. 5). Gas stream overpressure entered into the network through the middle pressure gas reduction station Z1 (Fig. 5) can be variable in range from 1700 to 2500 Pa. From the practical point of view, the overpressure of the gas entered to the network should be possible lowest, as this will allow to maintain respectively low gas pressure in the network, that in case of leak or the network failure will allow to minimize gas losses. Such effect can be achieved by steering appropriately gas stream pressure feeding the network in point Z1.

Steering gas stream pressure inputted into the network can be done with several methods. One of the methods is to connect pressure of gas stream p_{z1} and stream size Q_{feed}. For that purpose it is necessary to develop relationship $p_{z1} = f(Q_{feed})$. Another method is to make dependent entry stream pressure on gas pressure in selected point of the network. In this case it is necessary to develop gas network characteristic and appointing network nodes, where pressure is the lowest, as those nodes shall be most exposed to possible entry pressure changes.

6.1 Pressure steering algorithm

During gas flow in the network simulation, for each entry data set (gas stream leaving network), overpressure gas stream entered into the network was multiple time changed, and it allowed to indicate (from range of 1700 ÷ 2500 Pa) the lowest gas overpressure feeding the network. During research a clear dependence was noticed of the feeding overpressure stream on the size of the stream entering the network. The stream overpressure p_{z1} dependence on the stream size Q_{feed}, being sum of streams received in 108 nodes of the network is presented on Fig. 9. The figure presents 96 points, that characterise minimal gas overpressure, that at given network load Q_{feed}, ensure gas delivery to all 108 gas consumption nodes under overpressure higher than 1700 Pa. Data on the Fig. 9 presenting results received for each of 24 hours and 4 days of various air temperature were described using the equation

$$p_{z1} = 5 \cdot 10^{-4} Q_{feed}^2 + 8.23 \cdot 10^{-2} Q_{feed} + 1703 \qquad (15)$$

Equation (15) allows to estimate the value of minimal gas stream overpressure p_{z1} [Pa] feeding the network depending on size of that stream Q_{feed} [m³/h] with average relative error of ± 2 %. The equation (15) includes influence of both time of the day and air temperature on the size of the stream overpressure feeding the network. Air temperature influences on the decrease of gas streams leaving the network in 108 consumption nodes, and hence there is smaller total gas stream entering the network Q_{feed}. Influence of the time of the day on feeding pressure value is considered also in form of gas stream size leaving the network. Higher gas consumption from the network can be noticed during day hours, whereas lower during night hours. Exact entry overpressure value change p_{z1} in further hours of the exemplar winter day (t = -16 or -4 °C), summer day ($t \geq 18$ °C) or autumn or spring day ($t = 8$ °C) presents in Fig. 10. Analysing results presented in Fig. 10 one can see, that value of the stream minimal overpressure feeding the network depends on the time of the day and achieves higher values in hours of maximal gas consumption from the network.

The equation (15) called pressure steering algorithm can be used to program the reducer in gas reduction station (point Z1 on Fig. 5). However, determination of such an algorithm requires many time consuming calculations.

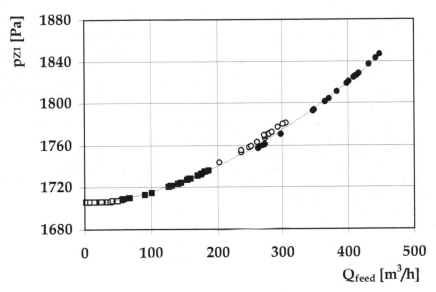

Fig. 9. The effect of gas stream size feeding the network Q_{feed} on the gas stream overpressure p_{Z1}; (●) t = -16 °C, (○) t = -4 °C, (■) t = 8 °C, (□) t ≥ 18 °C.

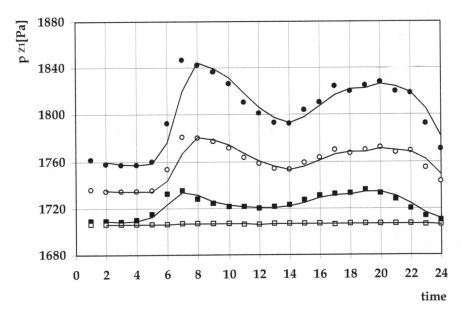

Fig. 10. The change of the gas stream overpressure p_{Z1} feeding the network in particular hours of the day; various values of air temperature; (●) t = -16 °C, (○) t = -4 °C, (■) t = 8 °C, (□) t ≥ 18 °C.

6.2 The characteristic of gas pipeline network

Based on gas flow in the network steady-state simulation results allowed to define also the network characteristic, that is presented in form of map of overpressure and gas streams layout in all gas pipeline of the network. Analysed data allowed to determine areas of the network, that are most sensitive to possible entry pressure stream fluctuation. Results presented in further part of the chapter were received for minimal value of the network feeding overpressure gas p_{z1}, but allowing to deliver gas to each node of the network under overpressure of $p \geq 1700$ Pa. Therefore on presented maps there are no areas with too low gas pressure. However based on detailed calculation results, received during simulation, one can at the same time define nodes of the network that due to their location in the network and characteristic of gas partition, are sensitive on gas pressure rapid decrease. Such nodes (called typical nodes) were marked on Figs. 11 - 14 with dashed line and their location in the network changes and depend on gas streams leaving the network. The reason of typical nodes location change in the network is irregular location of gas collection nodes, what causes uneven network load. In this case, even small gas pressure decrease at the network entry, below minimal value, will cause that gas streams consumed by the recipients focused in these nodes, will characterise with overpressure lower than minimal accepted ($p = 1700$ Pa), and this may cause damage or faulty running of appliances feeding with gas.

Joining typical node or nodes of the network with gas reduction station (point Z1) can be used in second mentioned gas pressure steering method. Gas pressure drop in typical node will be a signal to the reduction station to increase gas pressure entered to the network and vice versa.

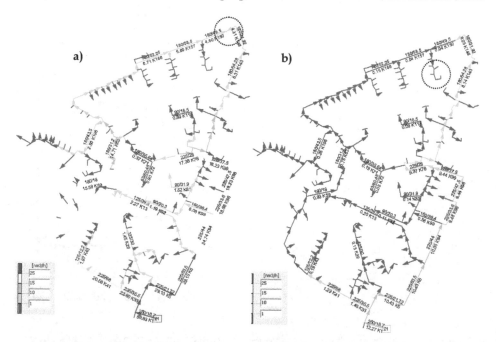

Fig. 11. Gas stream arrangement in pipelines network; t ≥ 18 °C; a) results for 7 am (p_{z1} = 1707 Pa; Q_{feed} = 55.83 m³/h); b) results for 3 am (p_{z1} = 1706 Pa; Q_{feed} = 12.27 m³/h).

The simulation results in form of gas streams layout in the gas pipeline network received for the hour of minimal and maximal gas consumption on the day of average air temperature higher than $t > 18$ °C are presented in Fig. 11. However, analogous data, received however for a winter day with average air temperature $t = -4$ °C presents Fig. 12. Comparing results presented in Figs. 11 and 12 one can see, that clearly higher load of the network is during winter in hours of top gas consumption (at 7 am), whereas lowest is in summer during night hours (at 3 am). In summer time in definitely larger part of gas pipelines, the stream of flowing gas is lower than 1 m³/h, whereas in winter is much higher and is approx. 8 m³/h.

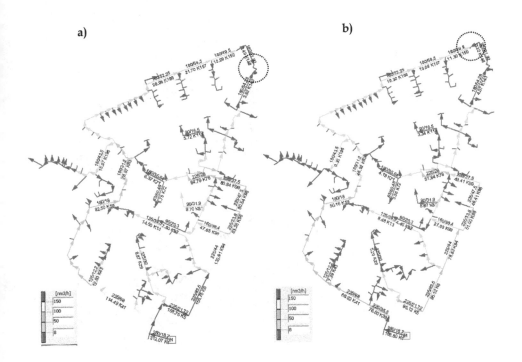

a)   b)

Fig. 12. Gas stream arrangement in pipelines network; $t = -4$ °C; a) results for 7 am (p_{z1} = 1781 Pa; Q_{feed} = 312.07 m³/h); b) results for 3 am (p_{z1} = 1734 Pa; Q_{feed} = 186.80 m³/h).

The network load increase caused by higher gas consumption from the network requires its feeding with stream of higher overpressure. Considering results presented on Fig. 13 and Fig. 14 gas overpressure layout can be analysed in all the network gas pipelines depending on the network load and feeding overpressure value. No significant differences were noticed in the overpressure values feeding the network in summer time (Fig. 13a and Fig. 13b). However in winter time (Fig. 14a and Fig. 14b) differences between gas overpressure values entered the network within hours of top consumption and during the night hours are clearly higher.

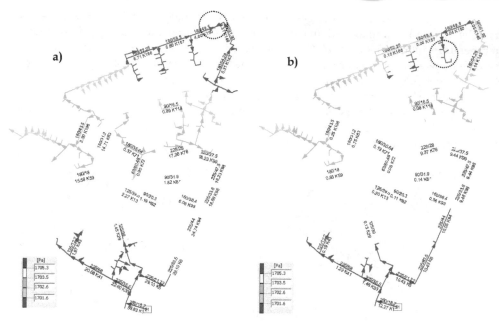

Fig. 13. Gas stream overpressure arrangement in pipelines network; $t \geq 18\,°C$; a) results for 7 am (p_{z1} = 1707 Pa; Q_{feed} = 55.83 m³/h); b) results for 3 am (p_{z1} = 1706 Pa; Q_{feed} = 12.27 m³/h).

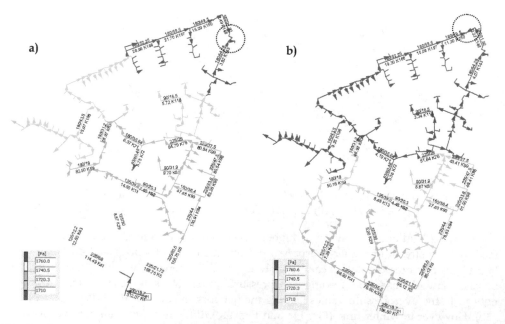

Fig. 14. Gas stream overpressure arrangement in pipelines network; t = -4 °C; a) results for 7 am (p_{z1} = 1781 Pa; Q_{feed} = 312.07 m³/h); b) results for 3 am (p_{z1} = 1734 Pa; Q_{feed} = 186.80 m³/h).

7. Conclusions

Natural gas transportation and distribution from production place is in gas pipelines that vary with gas overpressure value, gas speed in pipe, network pipeline materials and the network structure.

Gas transportation to consumers is a difficult task, because characteristic feature of gas flow in the network is irregularity of the network load caused with irregularity of gas consumption by consumers from the network, that shift in seasonal and daily cycle.

Gas transportation through network depends inter alia on the work quality of the gas network objects (compressor stations, reduction stations and gas pipelines). Large number of works dedicated to work optimisation of objects proves importance of this issue.

Assuming, that gas loss resulting from the network leak are proportional to gas pressure in the pipeline, steering gas pressure feeding the network allows to lower significantly the network exploitation costs due to lower gas loss, caused with the network leak or with the network gas pipeline damage.

8. Acknowledgment

The author would like to thank to the President and workers of the Gas Company in Szczecin, for access and preparation of data to necessary for preparation of the present work and for providing precious guidelines, enabling correct interpretation of the calculation results.

9. References

Brkić D. (2009). An improvement of Hardy Cross method on looped spatial natural gas distribution networks. *Applied Energy*, 86, pp. 1290-1300.

Carvalho A.A., Rebello J.M.A., Sagrilo L.V.S., Camerini C.S. & Miranda I.V.J. (2006). MFL signals and artificial neural networks applied to detection and classification of pipe weld defects. *NDT&E international*, 39, pp. 661-667.

Chaczykowski M. (2010). Transient flow in natural gas pipeline – The effect of pipeline thermal model. *Applied Mathematical Modelling*, 34, pp. 1051-1067.

Fukushima K., Maeshima R., Kinoshita A., Shiraishi H. & Koshijima I. (2000). Gas pipeline leak detection system using the online simulation method. *Computers & Chemical Engineering*, 24, pp. 453-456.

Herran-Gonzalez A., De La Cruz J.M., De Andres-Toro B., Risco-Martin J.L. (2009). Modeling and simulation of a gas distribution pipeline network. Applied Mathematical Modelling, 33, pp. 1584-1600.

Hwang K., Mandayan S., Udpa S.S., Udpa L., Lord W. & Atzal M. (2000). Characterization of gas pipeline inspection signals using wavelet basis function neural networks. *NDT&E international*, 33, pp. 531-545.

Ke S.L. & Ti H.C. (2000). Transient analysis of isothermal gas flow in pipeline network. *Chem. Eng. Jour.*, 76, pp. 169-177.

Kralik J., Stiegler P., Vostry Z. & Zavorka J. (1988). *Dynamic Modeling of Large-Scale Networks with Application to Gas Distribution*. Elsevier, Amsterdam

Liu M., Zang S. & Zhou D. (2005). Fast leak detection and location of gas pipeline based on an adaptive particle filter. *Int. J. Appl. Math. Comput. Sci.*, 15, pp. 541-550.

Nguyen H.H. & Chan Ch.W. (2006). Application ofartificial intelligence for optimization of compressor scheduling. *Engineering Applications of Artificial Intelligence*, 19, pp. 113-126.

Nguyen H.H., Uraikul V., Chan C.W. & Tontiwachwuthikul P. A. (2008). Comparison of automation for optimization of compressor scheduling. *Advances in Engineering Software*, 39; pp. 178-188.

Osiadacz A.J. & Chaczykowski M. (2001). Comparison of isothermal and non-isothermal pipeline gas flow models. *Chem. Eng. Jour.*, 81, pp. 41-51.

Osiadacz A.J. (1987). *Simulation and analysis of gas network*. E&FN Spon., London.

Osiadacz A.J. (2001). *Steady- state simulation of gas network*, BIG, Warszawa (in Polish).

Reddy H.P., Narasihman S. & Bhallamudi S.M. (2006). Simulation and state estimation of transient flow in gas pipeline networks using a transfer function model. *Ind. Eng. Chem. Res.*, 45, pp. 3853-3863.

Silva R.C.C., Guerreiro J.N.C. & Loula A.F.D. (2007). A study of pipe interacting corrosion defects using the FEM and neural networks. *Advances in Engineering Software*, 38, pp. 868-875.

Szoplik J. (2010a). Quantitative analysis of the heterogeneity for gas flow in the pipeline system. *Gaz, Woda i Technika Sanitarna*, 2010, 1, 2-6 (in Polish).

Szoplik J. (2010b). The application of the graph theory to the analysis of gas flow in a pipeline network. *37th International Conference of SSCHE*, proceedings on CD ROM, 24-28.05.2010, Tatranske Matliare, Slovakia

Szoplik J. (2010c). The steady-state simulations for gas flow in a pipeline network. *Chem. Eng. Trans.*, 21, pp. 1459-1464.

Tao W.Q. & Ti H.C. (1998). Transient analysis of gas pipeline network. *Chem. Eng. Jour.*, 69, pp. 47-52.

Zahedi G., Karami Z. & Yaghoobi H. (2009). Prediction of hydrate formation temperature by both statistical models and artificial neural network approaches. *Energy Conversion and Management*, 50, pp. 2052-2059.

Part 2

Natural Gas Utilization

4

Natural Gas Dual Reforming Catalyst and Process

Hamid Al-Megeren[1] and Tiancun Xiao[2,3,*]
[1]Petrochemical Research Institute,
King Abdulaziz City for Science and Technology, Riyadh,
[2]Inorganic Chemistry Laboratory, Oxford University,
[3]Guangzhou Boxenergy Technology Ltd, Guangzhou,
[1]Saudi Arabia
[2]UK
[3]PR China

1. Introduction

Natural gas consists primarily of methane, typically with 0–20% higher hydrocarbons. It is always associated with other hydrocarbon fuel, in coal beds, as methane clathrates, and is an important fuel source and a major feedstock for fertilizers. Natural gas is commercially extracted from oil fields and natural gas fields. Gas extracted from oil wells is called associated gas. The natural gas industry is extracting gas from increasingly more challenging resource types: sour gas, tight gas, shale gas, and coalbed methane. In fact, methane can also be produced from landfill site or anaerobic digestion, which is called biogas, it is in fact a kind of young age natural gas.

Components	(mol %)
CO_2	71
CH_4 + C_2+ hydrocarbons	28
H_2S	0.5
N_2	0.5

Table 1. Typical composition of Natuna gas field (Richardson and Paripatyadar 1990; Suhartanto, York et al. 2001)

In naturel gas, non-hydrocarbon gases (CO_2, N_2, H_2S) can account between 1% to 99% of overall composition. High carbon dioxide (CO_2) concentrations are encountered in diverse areas including South China Sea, Gulf of Thailand, Central European Pannonian basin, Australian Cooper-Eromanga basin, Colombian Putumayo basin, Ibleo platform, Sicily, Taranaki basin, New Zealand and North Sea South Viking Graben. The composition of CO_2 can reach as high as 80% in certain natural gas wells such as wells at the LaBarge reservoir in western Wyoming and the Natuna production field in Indonesia. Besides, purged gas from a

---

* Corresponding Author

gas-reinjected EOR (Enhanced Oil Recovery) well can contain as much as 90% CO_2 (Ricchiuto and Schoell 1988; Schoell 1995; Ballentine, Schoell et al. 2001; Cathles and Schoell 2007).

So far in industry, there have been several ways to bringing gas to the market. These include gas pipeline, liquid natural gas (LNG), and compressed natural gas (CNG). The costs of different fuel transportation are shown in Fig. 2. Also in gas transportation, any impurities such as CO_2 and H_2S must be fully removed, so as to protect the gas pipeline and container. This again will emit lot s of CO_2, the main green house gas into atmosphere. In terms of cost, as shown in Fig. 2, clearly the smallest gas field has highest transportation cost in both onshore and off shore. In a relatively high gas field, the transportation cost increase not as fast as the small gas field, but still much faster in transportation in cost, while the oil transport cost increases much smaller compared to the natural gas. Normally the gas and oil field are co-existed, and far away from the market, so the transportation cost must be taken into consideration. This may be another reason more and more attentions are being paid to the conversion of the gas into liquid, e.g., gas to liquid, and then the resultant liquid is mixed with the crude oil for transportation.

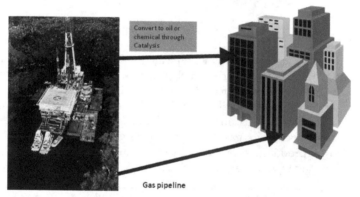

Fig. 1. The ways to bring natural gas into market

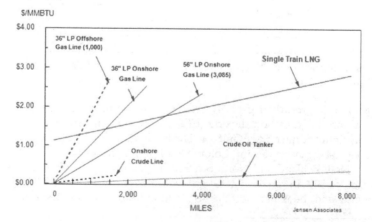

Fig. 2. The dependence of transportation cost on the distance for various fuels.

Indeed, Gas-to-liquids (GTL) has been a technology operated by industry for many years, while continuing improvement. It is a developing technology that converts stranded natural gas into synthetic gasoline, diesel or jet fuel through the Fischer-Tropsch process. Such fuel can be transported to users through conventional pipelines and tankers. Proponents claim GTL burns cleaner than comparable petroleum fuels. Most major international oil companies are in an advanced stage of GTL production, with a world-scale (140,000 barrels (22,000 m³) a day) GTL plant in Qatar already in production in 2009.

The gas to liquids (GTL) process comprises three main process steps, firstly the reforming of natural gas to syngas, a mixture of gases containing hydrogen, carbon monoxide, carbon dioxide and unreacted methane, secondly the Fischer Tropsch (FT) conversion of carbon monoxide and hydrogen to long chain hydrocarbons and thirdly the upgrading and refining of these hydrocarbons into a specific state of liquid fuels.. The syngas step converts the natural gas to hydrogen and carbon monoxide by partial oxidation, steam reforming or a combination of the two processes. The key variable is the hydrogen to carbon monoxide ratio with a 2:1 ratio recommended for F-T synthesis. Steam reforming is carried out in a fired heater with catalyst-filled tubes that produces a syngas with at least a 5:1 hydrogen to carbon monoxide ratio, as shown in Fig 3. To adjust the ratio, hydrogen can be removed by a membrane or pressure swing adsorption system. Helping economics is the surplus hydrogen is used in a petroleum refinery or for the manufacture of ammonia in an adjoining plant. Because of the steam reforming occurring under very harsh conditions and also giving much higher H_2/CO ratio, it accounts up to 60% of the GTL cost (Shamsi and Johnson 2001). Another way to give ideal H_2/CO mixture syngas is through partial oxidation, which has been well developed in lab scale and going to industrial pilot operation.

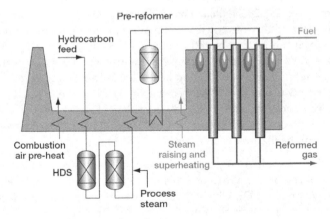

Fig. 3. Diagram of industrial steam reforming process

Another way to give approximate H_2/CO ratio of 2 is to combine dry and steam reforming, e.g., through the following reactions

$$CH_4+H_2O===CO+3H_2 \tag{1}$$

$$CH_4+CO_2===2CO+2H_2 \tag{2}$$

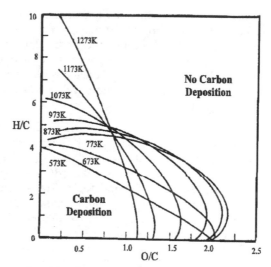

Fig. 4. The dependence of carbon deposition on H/C and O/C in the reforming system
(Shimura, Yagi et al. 2002; Zhang, Li et al. 2004)

This combining process here is named as dual reforming process, which can theoretically
give H_2/CO in the reaction products to 2:1 ratio, which is suitable for the follow up Fischer
Tropsch synthesis. Also this can make use of the CO_2 in the natural gas, especially in the gas
field where high CO_2 may be contained. It is well known that transporting the gas through
LNG or CNG or even pipeline transportation requires eliminating the CO_2 to avoid the
corrosion. (Butts 2006; Do 2007; Bhattacharya, Newell et al. 2009) Converting the CO_2
containing gas into liquid through the dual reforming process not only converts the natural
gas into liquid, which can be transported as liquid with low cost, but also can make use of
the CO_2 in the gas, thus to reduce green house gas emission. The dual reforming process in
principle does not require any modification of the steam reforming reactor, but may have
much harsh requirements on the catalyst, because the H_2/C in the products would be much
less than the steam reforming alone (Wang, Li et al. 1996; Ding, Yan et al. 2001; Shamsi and
Johnson 2003), also the O/C in the products is not very high, carbon thus very easily occurs
over the system from the thermodynamic view.

Recently steam/CO_2 methane reforming in the syngas reformer is one of particularities of
our new catalysts development. We intend to stick to the currently in-use reforming reactor,
but feeding both steam and CO_2, which already exists in the natural gas feedstock ranging
from 10% to 60%. Through properly adjusting the CO_2 and H_2O ratio, we hope to utilize
CO_2 contained in the natural gas and does not require any CO_2 separation. The Ni based
reforming catalyst has been prepared using a special dispersion agent, and an oxygen
storage material is introduced into the lattice of spinel support, so as to depress the carbon
formation. The catalyst has been successfully applied in steam reforming of natural gas for
ammonia industry, and the life time has been as long as 6 years. Here we report the
application of the catalyst in dual reforming for syngas production. The optimal operating
conditions for dual reforming have been explored.

2. Catalyst design

The preparation of steam reforming catalysts based on nickel as the active component has been extensively and intensively studied for years. Over a given nickel contained catalyst with similar nickel metal dispersion (which can be affected by the preparation, activation and operation conditions), the industrial catalyst activity would depends more in the catalyst shape. In fact, the major geometric factor affecting the activity of catalyst particles is the ratio of the particle's geometric surface area to volume, SA/V. When applying the catalyst to industrial reactor, the optimization of catalyst particles must take the three key factors, e.g., low pressure drop, high radical and high strength into consideration. These factors have direct relations with the shapes and size of the catalyst. For example, the high voidage e.g., large particles gives low pressure drop, however, smaller particles can give higher surface area, leading to more active site exposed to the reactants. Also due to the strong endothermic reaction properties, the reactor requires the catalyst to have good radial mixing so that the heat can be transferred from the reactor wall to the centre for the reaction. Also the strength of the catalyst particles are more important, due to the high pressure and high temperature conditions, the catalyst strength must be strong enough to resist the variations.

In the early stage of reforming catalyst development, ring or rasching shapes have been developed. Normally the smaller the particles the higher the catalytic activity and the better heat transfer properties but the higher the pressure drop as the gases pass through the reformer tube. Since 1990s, more and more attentions are been paid to catalyst geometric shapes to increase the geometrical surface area and hence the catalytic activity. The geometric surface area of a catalyst should be $200m^2/m^3$ or above. In recently years, strong shaped reforming catalysts with higher surface areas and voidage has been developed, as shown in Fig 5. Given the same dispersion of the active metal in the overall catalyst, which can give the same intrinsic activity when the mass transfer factors are eliminated, in industrial application, the physical properties of the catalyst in fact exert more influence on the catalyst performance. From Table 2, it is expected that the catalyst with 10 holes would have higher activity. This has also been confirmed by our industrial operation results.

Fig. 5. The novel shapes with high strength of the reforming catalyst and support.

Catalyst Geometry	Surface Area(m²)	Void Fraction	SA/V(m⁻¹)
1-hole	0.00503	0.66	1.151*10³
1-hole -6-grooves	0.0052	0.72	1.733*10³
4-hole	0.00523	0.62	1.703*10³
10-hole	0.00642	0.7	2.013*10³

Table 2. The physical properties of various shaped reforming catalysts.

3. Dual reforming catalyst performance test

3.1 Experiments

The reactions were carried out in a 70cm stainless steel reactor tube with id and od of 1.2cm and 2cm respectively. The furnace, comprising 6 individually heated zones, was specially designed to allow a temperature profile to be set across the reactor mimicking that of an industrial reformer. Four thermocouples were positioned inside a thermowell (inside the catalyst bed) to record the temperature of the catalyst bed at the inlet, exit and two intermediate positions. A diagram of the reactor and test setup is shown in Figure 6.

The rig has a supply of CH_4, CO_2, H_2, N_2 controlled by mass-flow controllers and an HPLC pump delivers H_2O, which is fed and mixed with the gases to flow together through a vaporiser where steam is generated before flowing into the reactor. Temperature programmes can be set on the vaporiser and all reactor furnace zones. Before any reactions were started, it was necessary to set a temperature profile across the reformer tube. The profile was designed to be similar to that of an industrial reformer, ensuring a cooler temperature at the tube inlet and gradually increasing the temperature towards the exit.

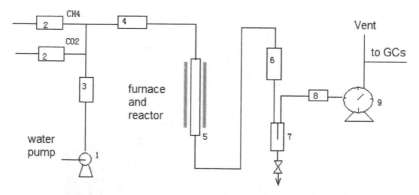

Fig. 6. Pressure Test system setup, 1–water pump, 2–gas MFC, 3–Purifier, 4–Mxier, 5–Tubular reactor, 6–condenser, 7–separator, 8–Back-Pressure regulator, 9–Wet flowing meter

In reforming reaction, carbon will most likely form via the methane cracking reaction which is promoted by high temperature:

$$CH_4 \Leftrightarrow C + 2H_2 \quad \Delta H = 75 \text{ KJmol}^{-1}$$

Figure 7 shows equilibrium data for CH_4 cracking with a CH_4 only feed at 3 bar.

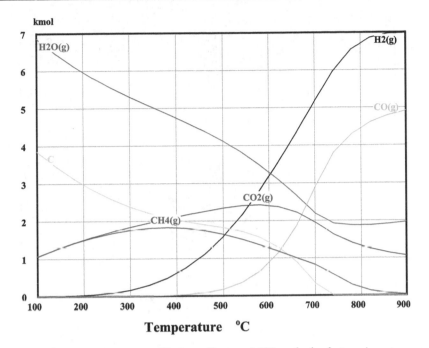

Fig. 7. CH₄ cracking reaction at equilibrium. 3bar and CH₄ only feed at various temperatures

At the catalyst bed inlet there is very little H_2 from reforming available to promote the reverse reaction and so the catalyst inlet is susceptible to coking by CH_4. It is therefore necessary to keep the temperature low enough where the thermodynamics of carbon formation via methane cracking are less favourable (Shamsi and Johnson 2001). However, it must also be sufficiently high to allow some H_2 production to prevent cracking further down the bed. Similar to an industrial reformer, the bed inlet temperature of the current study was 550°C (Valdes Perez, Fishtik et al. 1999; Shamsi and Johnson 2003; Snoeck, Froment et al. 2003).

Another common carbon forming reaction is the Boudouard reaction (Riensche and Fedders 1991):

$$2CO \Leftrightarrow C + CO_2 \quad \Delta H = -171 \ KJmol^{-1}$$

No carbon is predicted from the Boudouard reaction at the bed inlet under the conditions in the present study as there is only CO_2 and no CO.

Towards the exit of the catalyst bed, enough H_2 has been produced from reforming to prevent C from methane cracking reaction and so the temperature can be increased. The reforming reaction is endothermic and favoured by high temperatures so it is desirable to use as high a temperature as possible. Figure 8 shows equilibrium data for a dual reforming reaction at 3 bar with feed $CH_4:CO_2:H_2O$ 1:1:1.

Clearly the high temperature favors the CH_4 conversion and H_2 and CO yields (Treacy and Ross 2004; Maestri, Vlachos et al. 2009), however, it is difficult for CO_2 to be fully converted

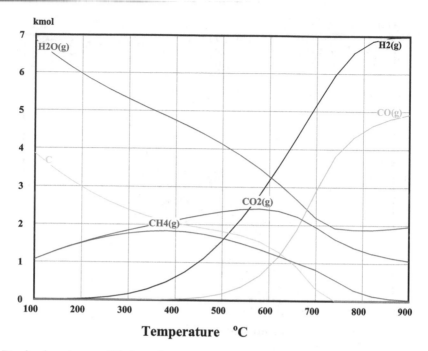

Fig. 8. Dual reforming equilibrium data at increasing temperature. 3bar, $CH_4:CO_2:H_2O$ 1:1:1 feed

into CO even at high temperature. Also at high reaction temperature, such as 750°C or above, no carbon deposition is expected from the thermodynamic views.

3.2 Catalyst charging

The test was carried out in a 100ml pressure reactor system to simulate industrial operation conditions. Natural gas and CO_2 were controlled to feed the reactor using two gas MFCs. De-ionised water was fed using a quantitative pump, first vaporisation then mixing with the gas feedstock. The mixing system was then heated to the required inlet temperature, and further going to pass the catalytic reactor. The reactor converter was a tubular style system, which was heated at different heating zones so as to mimic the industrial reactor temperature distribution. Thermocouples were installed inside the tube to monitor the catalyst bed temperature at different points. The products from the reactor were firstly cooled down to separate liquid, so as to become dry gas. The dry gas was measured and the composition was analysed. The schematic setup can be seen in Fig. 6.

3.3 Catalyst properties

The catalyst sample is from the industrial proved catalyst system for natural gas reforming, whose physical property is shown in table 3. This catalyst combination has been used in both China and abroad in the industries of ammonia, synthetic methanol, industrial hydrogen generation, and the performance is extraordinary.

Indicator	BOXE-1	BOXE-2
Shape and color	10-hole pellet, gray	1-hole grooved pellet gray
Dimensions (od×h) (mm)	Φ10×14	Φ14×14
density (kg/l)	1.10~1.15	1.10~1.15
Radius-strength (N/pellet)	>350	>500
Main composition NiO (m/m) support	>13 Al_2O_3	>13 $CaAl_2O_4$

The catalyst was reduced before use so as to convert nickel oxide into active metal.
$NiO+H_2=Ni+H_2O$

Table 3. The dual reforming BOXE catalyst properties

The physical properties of the steam reforming supports are as follows:

For the 10-hole pellet shaped catalysts, the support consists of pure α- Al_2O_3, surface area: 3.0~3.5m²/g; pore volume: 0.2~0.25ml/g;

For the 1-hole grooved pellet shaped catalyst, the main phase of the support is $CaAl_2O_4$ Surface area: 5~8m²/g; Pore volume: 0.2ml/g; average pore size: 1000A;

Pore distribution:

>1500A:	75.70%,
1000~1500A:	10.7%,
500~1000A:	8.58%,
100~500A:	3.85%,
<100A:	1.23%

3.4 Feedstock

Natural gas is civil use grade, whose composition is shown in table 4. CO_2 is industrial grade, and the de-ionised water is normal purified grade. The catalyst reduction hydrogen is industrial grade H_2.

component	Volume percentage %
methane	96.0
ethane	2.35
nitrogen	0.57
CO_2	1.08

Table 4. Typical low CO_2 containing natural gas composition

3.5 Catalyst activation

After catalyst is loaded, pressure test was carried out to ensure no leak occuring. Then the reactor temperature was increased at 10 °C/min to 500°C at inlet, 720°C in the middle and 800°C at the outlet, then hydrogen reduction was started and last for 8 hours

The reduction media is H_2/H_2O at ratio of 1:5, which is the same as industry. Hydrogen GHSV is about =1000h⁻¹ at ambient pressure.

4. Results and discussion

4.1 Test under pressure conditions

This series test is to mimic industrial practical process and operation. The conditions are under high pressure at $H_2O/C=2.0$ at low inlet temperature 500°C. These conditions are widely used in industry taking the techo-economical factors.

4.1.1 CO_2 fed amount effect on the product composition at $H_2O/C=2.0$

Pressure	2.6Mpa,
CH_4 GHSV	1200h⁻¹,
Inlet T:	500°C
Outlet T:	900°C

Pressure 2.6Mpa,
CH_4 GHSV 1200h^{-1},
Inlet T: 500°C
Outlet T: 900°C

The effect of CO_2 feeding (in terms of CO_2 GHSV) on the product composition is shown in Table 5. Each test was carried out for 400 hours to get steady and reproducible results. In industry, the catalyst performance in most case is measured in terms of H_2 and CO yield, and the reactants conversion. The conversion of methane is referred to the slip methane content, meaning the concentration of methane in the products (the dry gas base after steam removal and C elimination). The higher slip methane in the products suggests lower catalyst activity.

No.	NG GHSV H⁻¹	CO_2 h⁻¹	CO_2/CH_4	Inlet T	Outlet T	CH_4 %	CO %	CO_2 %	H_2 %	H_2/CO	CH_4 %
1	1200	2200	1.83	502	901	1.67	27.1	27.6	43.63	1.61	2.31
2	1200	2700	2.25	501	899	1.23	28.2	31.6	38.97	1.38	1.80
3	1200	3200	2.67	500	899	1.00	30.0	34.0	34.10	1.14	1.38
4	1200	3700	3.08	500	896	0.78	32.2	38.0	29.02	0.90	1.28
5	1200	4200	3.50	501	899	0.60	33.3	41.8	24.30	0.73	1.03

Table 5. Dependence of product composition on CO_2 feeding amount

The dependence of H_2/CO ratio on the CO_2/C ratio can be seen in Fig. 9, and the effect of CO_2/C on the slip methane content can be seen in Fig. 10.

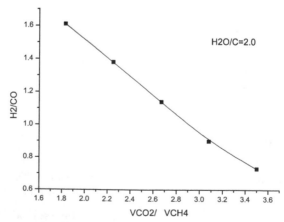

Fig. 9. Dependence of H_2/CO ratio in the product on the CO_2/C ratio in feedstock

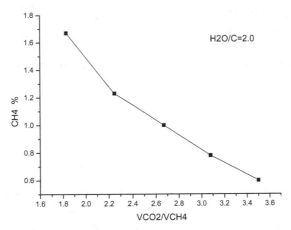

Fig. 10. Effect of CO_2/CH_4 ratio in the feedstock on slip CH_4 content

Clearly Fig 10 shows that the increase of CO_2 in the feedstock leads to the decrease of H_2/CO ratio in the products, and the slip methane content also decrease. This is helpful for the methane conversion and CO production.

4.1.2 Effect of pressure on the composition of products

The reaction conditions were set as follows;

H_2O/C:	2.0,
Methane GHSV:	1200h⁻¹,
CO_2 GHSV:	4200h⁻¹,
Inlet T:	500°C
Outlet T:	900°C

The effect of pressure on the gas product composition is shown in Table 6. Please note that each test was continued for 100 hours to get the steady state results, during the time on stream test, no catalyst activity drop or selectivity changes were observed, suggesting that the catalyst has high stability under the test conditions

No.	P MPa	Inlet T	Outlet T	CH₄ %	CO %	CO₂ %	H₂ %	H₂/CO	After CO₂ removal CH₄%
1	2.2	501	899	0.50	33.4	40.9	25.2	0.75	0.85
2	2.6	501	899	0.60	33.3	41.8	24.3	0.73	1.03
3	3.0	497	900	0.76	31.9	42.0	25.3	0.79	1.31

Table 6. Dependence of product composition on the pressures

With the increase of pressure, the slip methane content rises in the products, while H_2/CO ratio was little changed. The is in agreement with the thermodynamic equilibrium, because the reforming reaction gives more mole of gases than the reactants, increase the pressure leads to lower methane conversion(Friedmann, Burruss et al. 2003; Shamsi and Johnson

2003; Zhang, Li et al. 2004). However, the overall delta T to the equilibrium is about 5°C, which showed the super performance of the BOXE reforming catalysts.

4.1.3 Effect of outlet temperature on the composition of the product

Conditions:

H_2O/C: 2.0,
P: 2.6Mpa,
Carbon GHSV in terms of CH_4:   1200h^{-1},
CO_2 GHSV: 4200h^{-1},
Inlet T: 500°C

The outlet temperature effect on the product distribution is listed in table 7.

No.	Inlet T	Outlet T	CH_4 %	CO %	CO_2 %	H_2 %	H_2/CO	After CO_2 removal CH_4%
1	500	872	0.89	28.9	42.9	27.3	0.94	1.56
	499	872	0.89	29.0	42.5	27.6	0.95	1.55
2	501	899	0.60	33.3	41.8	24.3	0.73	1.03
	501	899	0.54	33.1	42.1	24.3	0.73	0.93

Table 7. The effect of outlet temperature on the gas product composition

Increasing outlet temperature favors the formation of CO, and helps to decrease the slip CH_4 content in the product. However, the outlet temperature may be restricted by the materials of the tubular reactor and energy consumption. The increase of outlet T may not be always a good solution to increase CH_4 conversion and decrease H_2/CO ratio. So far, most industrial operation for BOXE reforming catalysts uses 870°C as the outlet temperature.

4.1.4 Effect of CO_2 added in the feedstock on the gas product composition

Under of CH_4 GHSV of 1272h^{-1}, the results were shown in Table 8. It is shown that at VCO_2/VCH_4 ratio of 1.51, H_2/CO in the gas product is about 1.5, and slip CH_4 is only 0.1%. Comparison with Table 3 suggests that at the same GHSV of carbon (hydrocarbons), decreasing H_2O/C ratio and reducing P favour the CO generation and reducing CH_4 content.

P Mpa	GHSV h^{-1} CH_4	CO_2	CO_2/CH_4	H_2O/C	T °C Inlet	Outlet	Product dry gas composition % CH_4	CO	CO_2	H_2	H_2/CO	After CO_2 removal C_{CH4}
0.3	1272	1283	1.01	1.5	600	900	0.1	27.6	16.1	56.2	2.04	0.12
0.3	1272	1492	1.17	1.5	602	900	0.1	29.5	16.5	53.9	1.83	0.12
0.3	1272	1687	1.33	1.5	601	900	0.1	30.5	17.8	51.6	1.69	0.12
0.3	1272	1916	1.51	1.5	603	900	0.09	31.5	19.8	48.6	1.54	0.11
0.3	1272	2148	1.69	1.5	605	900	0.06	32.5	21.5	45.9	1.41	0.08
0.3	1272	2336	1.84	1.5	605	900	0.05	34.9	22.7	42.4	1.21	0.06

Table 8. Effect of CO_2 added in the feedstock on the product composition at relatively low pressure range

| P Mpa | GHSV h⁻¹ | | CO₂/C | H₂O/C | T ℃ | | Gas composition % | | | | | After CO₂ removal CH₄ (%) |
	CH₄	CO₂			Inlet	Outlet	CH₄	CO	CO₂	H₂	H₂/CO	
0.3	2561	1283	0.5	1.5	602	900	0.25	26.3	8.7	64.8	2.46	0.27
0.3	2561	2596	1.01	1.5	603	900	0.09	31.5	14.8	53.6	1.70	0.11
0.3	2561	2996	1.17	1.5	602	895	0.08	32.5	17.2	50.2	1.54	0.10
0.3	2561	3406	1.33	1.5	602	894	0.07	33.6	18.8	47.5	1.41	0.09
0.3	2561	3867	1.51	1.5	603	900	0.05	34.7	20.8	44.5	1.28	0.06

* Each test lasts for 400 hours to get steady and stable results

Table 9. Under relatively high methane GHSV, the dependence of gas product composition on the CO_2 feeding amount

Figs 10 and table 8 results suggest that low pressure and low H_2O/C ratio are favourable for methane conversion. Even when CH_4 GHSV was increased by one time (2561h⁻¹), slip methane in the gas products does not change significantly. However, the GHSV of methane has significant effect on the H_2/CO ratio. At CH_4 GHSV of 2561h⁻¹, and even when VCO_2/VCH_4 is only 1.17, H_2/CO is already nearly 1.5 in the product, suggesting high GHSV of methane is good for producing syngas with low H_2/CO ratio. This may be more useful for the production of syngas for carbonyl synthesis such as MEG production. It is contrary to the many industrial results.

The effect of GHSV of methane and CO_2/CH_4 ratios in the feedstock on the H_2/CO ratios are summarised in table 10. It is shown that the retention time exerts significant effect on the H_2/CO ratio in the products. High GHSV, e.g., shorter contact time of the reactants would have lower H_2/CO with the increase of VCO_2/VCH_4, which is much lower than the system with lower GHSV. The results are in agreement with the literature (Zhang, Li et al. 2004; Chen, Chiu et al. 2010). This suggests that the reaction is kinetics controlled, It may be worthwhile increasing the pressure so at the narrow the gap between the high and low GHSV for H_2/CO ratios.

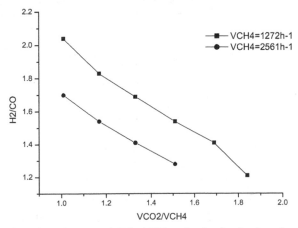

Fig. 11. Effect of GHSV of methane and CO_2/CH_4 ratios in the feedstock on the H_2/CO ratio in the products.

P Mpa	GHSV h⁻¹		CO_2/C	H_2O/C	T °C		Gas composition %					
	CH_4	CO_2			Inlet	Outlet	CH_4	CO	CO_2	H_2	H_2/CO	% CH_4*
0.3	2561	2592	1.01	1.3	610	900	0.13	32.4	13.7	53.8	1.66	0.15
0.3	2561	2592	1.01	1.5	601	900	0.08	31.6	14.8	53.6	1.70	0.11
0.3	2561	2592	1.01	2.0	600	899	0.06	27.6	17.3	55.0	1.99	0.07
0.3	2561	3406	1.33	1.5	602	894	0.07	33.6	18.8	47.5	1.41	0.09
0.3	2561	3406	1.33	2.0	601	900	0.04	30.5	22.9	53.4	1.75	0.05
0.3	2561	2592	1.01	1.5	601	900	0.08	31.6	14.8	53.6	1.70	0.11
0.45	2561	2592	1.01	1.5	603	895	0.35	29.5	15.4	54.8	1.86	0.41
0.6	2561	2592	1.01	1.5	608	890	0.65	29.2	15.7	54.5	1.87	0.77

* The CH_4 concentration refers to the one after excess CO_2 removal.

Table 10. Effect of CO_2 amount in the feedstock on the products compositions, CH_4 GHSV $= 2561h^{-1}$

4.1.5 The effect of CO_2 and H_2O added in the feedstock on the gas product composition under the constant CH_4 GHSV H_2O/C ratio, P and temperatures

The catalyst has been tested at various ratios of H_2O/CH_4 and CO_2/CH_4 at different pressures and outlet temperatures. The highest H_2/CO ratios with a low slip methane results were obtained under the conditions of 0.3Mpa, CO_2/CH_4 1.01 and H_2O/methane 2.0 at inlet temperature of 602°C and outlet temperatures of 899°C. The gas products can be used directly for Fischer Tropsch. However, if a CO rich products are desired, the reaction conditions can be changed to 0.3Mpa, CH_4 GHSV 2561h⁻¹, CO_2 GHSV 3406h⁻¹, inlet temperature of 602°C and outlet temperature of 894°C, under which H_2/CO is 1.41 and slip methane of 0.09.%.

The results in Table 10 also suggests that decreasing H_2O/methane ratio to as low as 1.3 with co-feeding CO_2 can effectively depress the carbon formation, and give a stable catalyst performance. This can significantly reduce the steam feeding in the conventional steam reforming process, which would save energy and simplify the hydrogen plant, also it can make use of CO_2 in the gas stream.

It is seen from Table 10 that when fixing the reaction pressure and CO_2/CH_4 ratio while increasing H_2O/methane ratios, H_2 and H_2/CO increase in the products. Increasing CO_2/CH_4 ratio favors the CO production, giving desirable product compositions for the process where more or pure CO is required.

When the pressure was increased to 0.6Mpa from 0.3Mpa, the slip CH_4 concentration increased more significantly. This can be explained by the thermodynamic model, due to the reforming reaction is volume expansion, while CO WGS is identical volume process (Steinfeld, Kuhn et al. 1993; Tsai and Wang 2008; Maestri, Vlachos et al. 2009).

4.1.6 Effect of lower H_2O/methane ratio on the reactions

At P of 0.3Mpa，and CH_4 GHSV of 2561h⁻¹，CO_2/CH_4 of 1 when decreasing H_2O/methane to as low as 1.0 and 0.8, the catalyst has run for more than 50 hours each，a little carbon deposition were observed in the inlet of the catalyst at H_2O/C at 0.8, but much more carbon deposition in the exit reactor tube has been observed, which caused big pressure drop. This suggests that the steam is very important and essential to depress the carbon formation. A further experiment also proves this. When the pressure drop increased to 2 bars during the reaction with H_2O/CH_4 0.8 for 60hours, the steam feeding was increased to H_2O/CH_4 at 2 with the CH_4 and CO_2 feeding rate unchanging, we can see that the H_2/CO ratio increased to 1.92, and slip methane decreased to 0.05%, also the pressure drop decreased from 2 bar to 0.1bar after 95 hours operation. This suggests that the deposited carbon in the catalyst bed can be removed through adjusting the H_2O/methane ratios.

5. Catalyst deactivation and regeneration study

During the catalyst pilot test, it is found that when the CO_2 is co-fed with the steam at $H_2O/CH_4>1$, the catalyst can effectively convert natural gas and CO_2 into syngas with various H_2/CO ratio, and no pressure drops were observed during the test. However, when the H_2O/CH_4 is decreased to less than 1, pressure drop occurs in the reactor, and increased with the time on stream, but the products distribution and methane conversion remained unchanged. This suggests that even with some carbon deposition, the catalyst active sites can still catalyze the reactions.

Here we collected the carbon deposited catalyst sample and characterized them using XRD, TG and laser Raman. The XRD results showed that nickel is present in highly dispersed metallic metal, while a small amount of crystalline carbon is detected, which show a very small diffraction peak at 28.9°, which is tentatively assigned to the carbon deposition.

The used catalyst has been measured using TG-DTA-MS, the weight change, heat flow and the emitted products were monitored using MS during the thermal analysis. For comparison, a fresh H_2/H_2O reduced Ni catalyst has also been characterised using TG-DTA-MS. The results are shown in Fig. 12. It is clearly shown that with the increase of

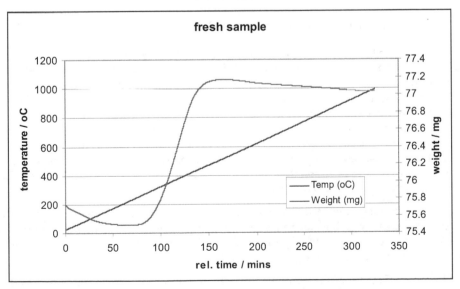

Fig. 12. TG-DTA-MS of the fresh reduced Ni/Al_2O_3 reforming catalyst

temperature, the weight of catalyst sample increase, which can be explained by the oxidation of nickel metal into nickel oxide. (Bhattacharyya and Chang 1994; Gonzalez-Delacruz, Pereniguez et al. 2011).

$$Ni+1/2O_2===NiO$$

Because the nickel metal particles uniformly dispersed over the support, the oxidation occurs in a very narrow range of the temperature from 230°C to 400°C. The weight loss after 500°C may be due to the decomposition of the resultant NiO.

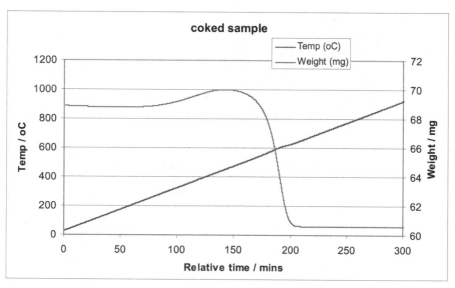

Fig. 13. TG-DTA of the used BOXE reforming catalyst

TG results of the used catalyst (unloaded from the reforming reactor where H_2O/methane=0.5 and operated for 50 hours), It is interesting to see that the catalyst weight started to increase at 210°C until 405°C, this is in agreement with the fresh catalyst sample, which can be explained by the oxidation of nickel metal particles into oxide. When the temperature is increased to 450°C, an abrupt weightloss occurs, with the weight of the sample change from 70mg to 60.5mg, equivalent the weighloss ratio of 13.8wt%. (Riensche and Fedders 1991; Claridge, Green et al. 1994; Verykios 2003; Adachi, Ahmed et al. 2009) Because the TG is carried in static air where oxygen is present, hence the deposited carbon is converted into CO_2 and evolved into the gas products, thus leading to the weightloss. This is further supported by the TG-DTA results as shown in Fig.14.

The TG-DTA results showed that the weighloss of the catalyst corresponds to a strong exothermic reaction, which is due to the oxidation of the carbon, as oxidation reaction gives heat. Also the carbon formation may be present in different forms, so the oxidation reaction is in a broad temperature ranges.

The MS results (Fig. 15) of the burnt gas suggest that only CO_2 is detected, which is in the same temperature for the weighloss of the catalyst sample as shown in Fig 13. No other gas

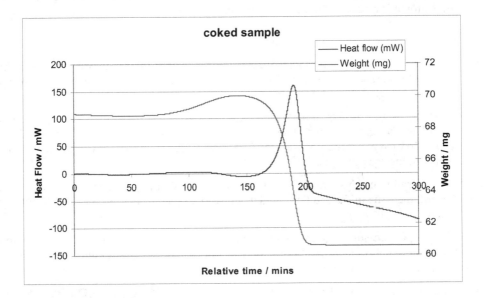

Fig. 14. TG-DTA result of the used BOXE reforming catalysts

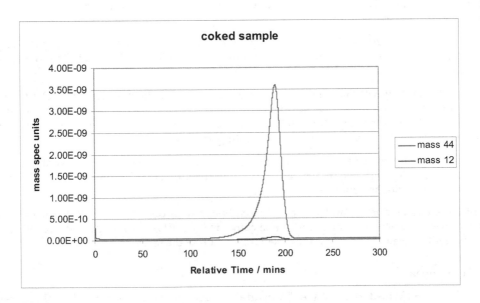

Fig. 15. The evolution of CO_2 from the used reforming catalysts during the thermal analysis

like methane or ethane is detected from the MS, also no aromatics, suggesting that the carbon formation over the BOXE catalyst is though nickel active site, not through the acid site, where aromatic carbon is the coke precursors.

6. Conclusion

6.1 Carbon dioxide is present in most natural gas field, separating CO_2 and transporting the natural gas into market may be more suitable for large gas field, but not economical for small gas field or high CO_2 containing gas field.

6.2 Converting the gas into liquid and transportation together with oil maybe an economical way to utilize the gas, but the conventional steam reforming catalyst requires high H_2O/CH_4 ratio to avoid carbon deposition, however, the resultant syngas has much higher H_2/CO ratio in the gas products.

6.3 BOXE reforming catalysts have been tested in dual reforming reaction to make use of both CO_2 and methane in the gas field, which showed a broad operation window for utilize the CH_4 and CO_2.

6.4 Under proper pressure, H_2O/C 、 CO_2/CH_4 ratios, reforming CH_4 using both H_2O and CO_2 at high temperature to generate syngas with low H_2/CO ratio is feasible. The catalysts showed stable good performance.

6.5 At temperature from inlet of 600°C and outlet of 900°C, given $H_2O/C=1.5$ to get $H_2/CO=1.5$ in the gas products is easy to obtain through adjusting GHSV and pressure and adding the amount of CO_2.

6.6 Reforming CH_4 at H_2O/C as low as 1.0 in the feedstock experience carbon deposition, especially in the reactor wall end. Hence it is not recommended to use the feedstock with low H_2O/C ratio in feedstock.

6.7 To reach long-term stable industrial operation, it is recommended to adopt relatively mild conditions such as lower outlet temperature (<870ºC) relative high H_2O/C ratio (ca.1.4-1.5) to do the pilot test.

6.8 Dual reforming with CO_2 and H_2O not only helps to give the desirable H_2/CO ratio for the reformed gas products, but also helps to make use of the CO_2 in the gas fields and reduces GHG emission, The BOXE reforming catalyst showed high activity and selectivity to the syngas production at a very broad operating window.

6.9 The low H_2O/methane feed may result in carbon deposition when CO_2/CH_4 ratio is not high enough. However, the results carbon can be converted into CO through increasing the H_2O feeding during the operation.

6.10 The carbon formation under lean steam/carbon ratio is due to the nickel decomposition of the methane over the catalysts, it can be converted into CO_2 when oxygen is present, this can be another way for the catalyst regeneration outside of the reactor.

7. References

Adachi, H., S. Ahmed, et al. (2009). "A natural-gas fuel processor for a residential fuel cell system." *J. Power Sources* 188(1): 244-255.

Ballentine, C. J., M. Schoell, et al. (2001). "300-Myr-old magmatic CO2 in natural gas reservoirs of the west Texas Permian basin." *Nature* 409(6818): 327-331.

Bhattacharya, S., K. D. Newell, et al. (2009). "Held tests prove microscale NRU to upgrade low-btu gas." *Oil Gas J.* 107(40): 44-48, 50-52.

Bhattacharyya, A. and V. W. Chang (1994). "CO2 reforming of methane to syngas: Deactivation behavior of nickel aluminate spinel catalysts." *Stud. Surf. Sci. Catal.* 88(Catalyst Deactivation 1994): 207-213.

Butts, R. C. (2006). "Processing low BTU gas from the Permian Basin Yates formation." *Annu. Conv. Proc. - Gas Process. Assoc.* 85th: butts clark1/1-butts clark1/10.

Cathles, L. M. and M. Schoell (2007). "Modeling CO2 generation, migration, and titration in sedimentary basins." *Geofluids* 7(4): 441-450.

Chen, W.-H., T.-W. Chiu, et al. (2010). "Enhancement effect of heat recovery on hydrogen production from catalytic partial oxidation of methane." *Int. J. Hydrogen Energy* 35(14): 7427-7440.

Claridge, J. B., M. L. H. Green, et al. (1994). "Methane conversion to synthesis gas by partial oxidation and dry reforming over rhenium catalysts." *Catal. Today* 21(2-3): 455-460.

Ding, R., Z. Yan, et al. (2001). "A review of dry reforming of methane over various catalysts." *J. Nat. Gas Chem.* 10(3): 237-255.

Do, N. T. (2007). "A new approach to utilize associated gas in the upstream business." *Sekiyu Gijutsu Kyokaishi* 72(2): 178-187.

Friedmann, J., R. C. Burruss, et al. (2003). "High CO2 gas fields: Natural analogs to the chemical evolution of CO2 storage in depleted gas fields." *Abstracts of Papers, 226th ACS National Meeting, New York, NY, United States, September 7-11, 2003*: GEOC-140.

Gonzalez-Delacruz, V. M., R. Pereniguez, et al. (2011). "Modifying the Size of Nickel Metallic Particles by H2/CO Treatment in Ni/ZrO2 Methane Dry Reforming Catalysts." *ACS Catal.* 1(2): 82-88.

Maestri, M., D. G. Vlachos, et al. (2009). "A C1 microkinetic model for methane conversion to syngas on Rh/Al2O3." *AIChE J.* 55(4): 993-1008.

Ricchiuto, T. and M. Schoell (1988). "Origin of natural gases in the Apulian Basin in south Italy: a case history of mixing of gases of deep and shallow origin." *Org. Geochem.* 13(1-3): 311-318.

Richardson, J. T. and S. A. Paripatyadar (1990). "Carbon dioxide reforming of methane with supported rhodium." *Appl. Catal.* 61(2): 293-309.

Riensche, E. and H. Fedders (1991). "Conversion of natural gas into carbon monoxide-rich syngases." *Stud. Surf. Sci. Catal.* 61(Nat. Gas Convers.): 541-547.

Schoell, M. (1995). "Geochemical characterization of CO2 containing natural gases in south east asia." *Book of Abstracts, 210th ACS National Meeting, Chicago, IL, August 20-24*(Pt. 1): GEOC-018.

Shamsi, A. and C. D. Johnson (2001). "Carbon deposition on methane dry reforming catalysts at higher pressures." *Abstracts of Papers, 221st ACS National Meeting, San Diego, CA, United States, April 1-5, 2001*: FUEL-049.

Shamsi, A. and C. D. Johnson (2003). "Effect of pressure on the carbon deposition route in CO2 reforming of 13CH4." *Catal. Today* 84(1-2): 17-25.

Shimura, M., F. Yagi, et al. (2002). "Development of a new H2O/CO2 reforming catalyst and process for natural gas utilization." *Prepr. - Am. Chem. Soc., Div. Pet. Chem.* 47(4): 363-365.

Snoeck, J.-W., G. Froment, et al. (2003). "Kinetic evaluation of carbon formation in steam/CO2-natural gas reformers. Influence of the catalyst activity and alkalinity." *Int. J. Chem. React. Eng.* 1: No pp given.

Steinfeld, A., P. Kuhn, et al. (1993). "High-temperature solar thermochemistry: production of iron and synthesis gas by iron oxide (Fe3O4) reduction with methane." *Energy (Oxford)* 18(3): 239-249.

Suhartanto, T., A. P. E. York, et al. (2001). "Potential utilisation of Indonesia's Natuna natural gas field via methane dry reforming to synthesis gas." *Catal. Lett.* 71(1-2): 49-54.

Treacy, D. and J. R. H. Ross (2004). "The potential of the CO2 reforming of CH4 as a method of CO2 mitigation. A thermodynamic study." *Prepr. Symp. - Am. Chem. Soc., Div. Fuel Chem.* 49(1): 126-127.

Tsai, H.-L. and C.-S. Wang (2008). "Thermodynamic equilibrium prediction for natural gas dry reforming in thermal plasma reformer." *J. Chin. Inst. Eng.* 31(5): 891-896.

Valdes-Perez, R. E., I. Fishtik, et al. (1999). "Predictions of activity patterns for methane reforming based on combinatorial pathway generation and energetics." *Prepr. Symp. - Am. Chem. Soc., Div. Fuel Chem.* 44(3): 541-545.

Verykios, X. E. (2003). "Catalytic dry reforming of natural gas for the production of chemicals and hydrogen." *Int. J. Hydrogen Energy* 28(10): 1045-1063.

Wang, S., G. Q. Li, et al. (1996). "Carbon Dioxide Reforming of Methane To Produce Synthesis Gas over Metal-Supported Catalysts: State of the Art." *Energy Fuels* 10(4): 896-904.

Zhang, Q.-H., Y. Li, et al. (2004). "Reforming of methane and coalbed methane over nanocomposite Ni/ZrO2 catalyst." *Catal. Today* 98(4): 601-605.

5

Innovative Technologies for Natural Gas Utilization in Power Generation

Victor M. Maslennikov, Vyacheslav M. Batenin,
Yury A. Vyskubenko and Victor Ja. Shterenberg
Joint Institute for High Temperatures of the Russian Academy of Sciences,
Russian Federation

1. Introduction

Numerous publications are dedicated to the development of traditional power engineering and to the possibility for essentially new power generation methods to appear. But if we shell a try to make detailed analysis of this issue for near perspective, the imagination has to be limited substantially. First of all, these limitations are connected with the great scale of the problem.

Appearance of a new technology in field of power generation, apparently may be considered as real, only when the first demonstration plant will comes into service or when the way of the transition from ideas and experiments to creation of the new plant is clear , together with needed time and resources. According to the progress in the last 40 years, a few of the principled new power production technologies have been realized [1,3]. Here, gas-turbine and combined cycle steam-gas technologies are the best examples. For the most parts progress consists of step-by-step enhancement and parameter increase, that had resulted in the qualitative growth of technical and economic indicators. In general, power engineering remains a very conservative field.

The second boundary condition of the origination of a new power production technology is the presence of the necessary volume of primary resources. However, the excess some of these resources can slow down the development of new technologies, based on the other primary resources.

Finally, with the start of XXI century, power generation development is more and more coordinated by the world community, which connects the increase of fossil based power production with the possible global climate change.

In respect to Russia, these boundary conditions are added by internal problems of new economic structure formation, which take rather long period of solution, according to the experience of the last two decades.

The estimations of organic fuel resources give their inevitable depletion in the foreseeable future. However, the exact figures are under permanent correction. Crude oil resources, apparently, are minimal even with taking into account the volumes which are difficult to extract and the development of new methods of extractions. Concerning natural gas, the estimations are more optimistic. The resources of natural gas assured and perspective for recovery grow (methane of coal beds and shale gas, for example). With respect to possibility of gas hydrates utilization, the reserves of gas seem to be practically unlimited for power production in the oncoming century.

The above mentioned is also true for the reserves of coal. Thereby, natural gas and coal are the most important primary energy resources.

With this, analyzing the perspectives of all power production technologies based on fossil fuels, it is worth taking into account the following:

- The problem of transportation is very actual because the gas and coal fields in most cases are distant from the areas of electric energy consumption;
- Transformation efficiency increase connected not only with fuel saving, but also with the reduction of greenhouse gases emission;
- Mono-production of electric energy does not correspond with long-term energy strategy;
- Even at keeping the present level of electric power production in Russia up to 2030, to compensate the retirement of power of thermal power plants only, it is necessary to put new 7 – 8 million of Kw/year in operation, where 4 – 5 million of Kw/year using natural gas, mainly at high capacity steam-gas plants.

Taking into account that not less than 10 years are to pass from draft technical proposal of new power production technology to the creation of big-scale demonstration plant, the problem of advance thermal efficiency at power production, natural gas first of all, should be solved in two directions:

1. By the upgrade at existing power plants, as a first stage;
2. The creation of principally new high efficient ecologically clean power units to replace the worked-out power units and for further development of fossil fuel power production.

Below is the description of several new technologies for natural gas utilization developed in Joint Institute for High Temperatures of Russian Academy of Sciences (JIHT RAS) in both areas.

2. Efficiency increase of existing power production facilities

2.1 Repowering of the existing water-heating boiler

Repowering of the existing gas fueled water-heating boiler by adding of gas-turbine units with its outlet to the boiler combustor for the first time was proposed by group of JIHT RAS specialists headed by Academician Mikhail Styrikovich in the 80th. At present this proposal is recognized and accepted for implementation, however big discrepancy exists in the reading of its technical essence, up to absurd (creation of gas-turbine units with exhaust-heat boilers at boiler facilities). The essence of the proposal is in the following: products of combustion from gas-turbine unit, containing 17% of free oxygen are fed to boiler combustor (Fig.1).

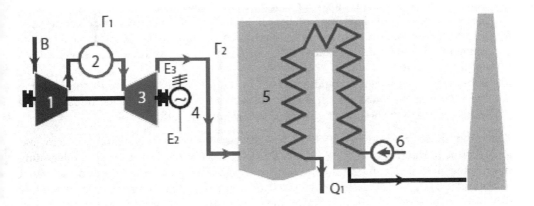

B – air, Г1 – gas to GT combustor, Г2   gas to boiler, 1 – compressor, 2 – combustion chamber, 3 – gas turbine, 4 – generator, 5 – boiler, 6 – feed pump.

Fig. 1. Water boiler with gas-turbine buildup.

With this, at minimal thermal load regime (20%) the heat is generated due to the cooling of the gas turbine unit products of combustion. At the increase of thermal load (up to nominal), additional fuel is after-burned in the boiler in the flow of products of combustion due to free oxygen combustion. With this, the efficiency of electric power production changes from 60% (summer regime) to 90% (winter regime). The efficiency is defined as:

$$\eta_e = \frac{N_e}{H_T - \dfrac{Q_m}{\eta_b}} \tag{1}$$

Water heating boilers of 100 Gcal/h capacity could be built-up by gas-turbine units of 16-20 MW (defined by the products of combustion flowrate). Only in Moscow at district heating units there are about 100 of such boilers.

The described technology has one peculiarity, namely, the gas flow rate to boiler combustor changes in a wide range at near constant flow rate of oxidant, represented by the gas-turbine products of combustion. This can be implemented by step-by-step some burners switch off or by the installation of special burners. The technology can be applied without additional research at test units and the cost of electricity generated is lower than at perspective steam-gas plants.

2.2 Repowering of the existing gas-fired steam turbine units

There are some variants to retrofit the steam turbine units with using gas turbines. A few obstacles need to be overcome for these possibilities to be namely:

1. the extent, to which the equipment of the power plant to be retrofitted is worn out, needs to be taken into consideration, so that its remaining service life would not differ too much from the service life of the newly installed equipment;
2. free space must be available for installing new equipment at the existing power plant;
3. it is necessary that the newly installed equipment should not bring about a substantial reduction in the capacity and efficiency of the existing basic power-generating equipment; and
4. the new process scheme should not result in the loss of reliability of the object being retrofitted; in the worst case, it should not increase the environmental impact on the region, while in the best case it should considerably relieve this impact.

The technical offer made by OIVTAN was aimed at solving this actual problem and involved the retrofitting of existing, relatively new steam turbine units by way of integrating gas turbines into these units and employing an original technology referred to as "partial oxidation" [2], [4], [5]. This technology essentially consists in that the natural gas, utilized by the steam-turbine unit, is preconverted in the combustors of the gas-turbine unit to carbon monoxide and hydrogen, and the resultant fuel gas expands in the gas turbine and then burned in the boiler furnace.

2.2.1 List of the units under consideration

The investigation, presented in this chapter, deals with a comparative feasibility study of the most representative alternative options. The main items under comparison are:

*   thermal efficiency additional power production,
*   relative cost of electricity in base mode operation.

Five alternative options have been studied:

1. A conventional 315 MW condensing steam-turbine unit fired with gas and fuel oil, for the steam parameters of 24 MPa and 540/540 oC. (It was used as a standard for comparison).
2. A 150 MW binary-cycle steam-gas unit (SGU) incorporating two W251B12 gas-turbine units by Westinghouse and a 51 MW steam turbine.
3. Gas-turbine topping with partial oxidation process.
    a. 7 variants with Aircraft Engines.
    b. 4 variants with Heavy-duty GTU.
4. Topping with dumping of GTU gas into the boiler furnace (Hot-Windbox).
    a. 4 variants with Aircraft Engines.
    b. 2 variants with Heavy-duty GTU.
5. Topping with GTU combined with STU in a single steam-generating circuit.

Standard steam-turbine unit

Units of this type were commissioned widely since the 1970s. Following the period of debugging, they have exhibited high reliability and efficiency and are at present employed most extensively in the energy systems of Russia and republics of the former USSR. The present-day characteristics of this unit are listed in the following table 1.

Generator power, MW	317.9
Steam parameters:	
-At turbine inlet	
Flow rate, kg/s	276.7
Pressure, kg/cm²	240.0
Temperature, °C	540.0
-At reheater outlet	
Flow rate, kg/s	221.3
Pressure, kg/cm²	38.6
Temperature, °C	540.0
-At condenser inlet	
Flow rate, kg/s	169.7
Pressure, kg/cm²	0.035
3. Feedwater temperature, °C	280.8
4. Pressure in deaerator, kg/cm²	7.0
5. Number of regenerative feedwater heaters	
- low pressure	4
- high pressure	3
6. Feedpump turbodrive power, MW	11.9
7. Specific heat consumption, KJ/KWh	8055
8. Efficiency of boiler , %	94.5
9. Unit net efficiency, %	40.6

Table 1. The main characteristics of standard steam-turbine unit.

2.2.2 An existing aircraft engine used as a topping unit in partial oxidation scheme

The schematic diagram is given in Fig. 2.

The existing engine in this particular case is used as a "gas generator". The combustion products containing up to 17% free oxygen are passed from the gas turbine exhaust to a special converter (11), where natural gas is fed in excess.

In the converter (11), natural gas at a temperature of about 1100 oC (possibly in the presence of a catalyst), is subjected to partial oxidation to hydrogen and carbon monoxide. The products of partial oxidation are expanded in the power gas turbine (3) and discharged as fuel to the upper tiers of the steam boiler. (The lower tier of the boiler may be operated on the former fuel).

Steam extracted from the STU exhaust to the GTU can be used in the topping for the following purposes:

- Cooling the power gas turbine blades;
- Increasing power output of the gas turbine (being fed into the combustion chamber of the aircraft engine (2);
- Decreasing the soot formation (being fed into the converter(11).

In order to enhance the efficiency some part of the combustion products may by-pass and fed to the additionally installed boiler feedwater preheaters. In doing so, the steam extracted from the steam turbine exhaust for feedwater preheating is changed to optimize thermal circuit.

The main characteristics of two variants of this unit are listed in the table 2.

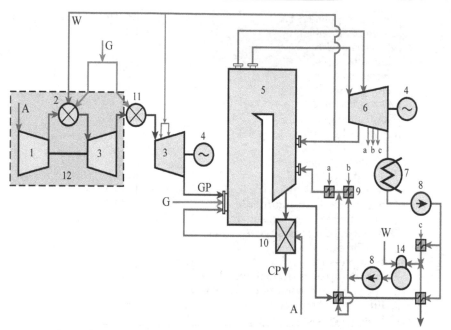

1 – compressor, 2 - combustion chamber, 3 - gas turbine, 4 - electric generator, 5 - steam boiler, 6 -steam turbine, 7 – steam condenser, 8 - feed pump, 9 - steam regeneration, 10 - air heater, 11 – partial oxidation chamber, 12 - gas generator (AGTE block), 13 - deaerator.

A -air, G -natural gas, CP - combustion products, GP - natural gas conversion products, W - water, steam

Fig. 2. Repowering by addition of topping SGU based on aircraft gas turbine with using partial oxidation.

Steam injection into GTU, kg/s	0.0		19.0	
MODES	1	2	1	2
Main characteristics				
1. GTU total power, MW	38.2	38.2	74.5	74.5
2. Additional gas flow to the boiler, kg/s	7.78	7.35	8.95	8.37
as % of the total flow	48.1	46.7	53.2	51.5
3. Efficiency of boiler , %				
- boiler per se	91.9	91.7	89.9	89.7
- system comprising boiler+gas regenerator	94.2	94.1	93.4	93.3
4. SGU total power, MW	320.4	311.1	300.4	288.8
5. Parasitics, MW	9.6	9.6	9.6	9.6
6. Additional useful power of the unit, MW	41.1	31.8	57.4	45.9
7. Total fuel consumption by the unit, kg/s	16.18	15.76	16.83	16.25
8. Efficiency of additional el. power production, %	69.0	82.6	62.5	72.8

Mode 1 - With gas regeneration. **Ne add = Nmax.**
Mode 2 - With gas regeneration. **Efficienci = Max.**

Table 2. The main characteristics of existing Aircraft Engine used as a topping unit in partial oxidation scheme.

2.2.3 A gas turbine with discharge of combustion products into the boiler furnace for repowering

This scheme (Fig. 3) is well established and it is mentioned here as alternative, the comparison being carried out for the same initial assumptions. In distinction from fig 1 addition topping gas turbine to steam generation block.

The combustion products containing up to 17% free oxygen are discharged to the boiler burners where they are used as an oxidizer of the additionally fed fuel. To control the boiler steam production rate, additional amounts of air and fuel are added to some burners.

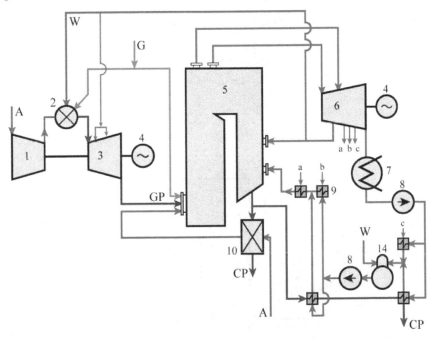

1 – compressor, 2 - combustion chamber, 3 - gas turbine, 4 - electric generator, 5 - steam boiler, 6 -steam turbine, 7 – steam condenser, 8 - feed pump, 9 - steam regeneration, 10 - air heater, 11 – deaerator.
A -air, G -natural gas, CP - combustion products, GP - natural gas conversion products, W - water, steam

Fig. 3. Repowering by addition of topping SGU, based on Westinghouse GTU.

Since the air flow through the air preheater (10) decreases significantly, part of the combustion products is removed via a by-pass for heating feedwater in parallel with the steam regeneration system. The flow rate of the by-passed combustion products in this unit is much higher, then in units based on partial oxidation and, by consequence, a larger fraction of the steam regeneration is forced out, which leads either to a loss of efficiency, or to a decrease in power output of the steam turbine. In the comparative thermodynamic analysis given hereafter we shall analyze the effect of amount of the combustion products discharged into the boiler downstream of the GTU as well as changes in the steam regeneration system on the efficiency of additional production of electric power and change in the steam turbine power output.

2.2.4 GTU integrated with Steam Turbine Unit in a single steam generating circuit

The schematic diagram of this option is given in Fig. 4. In this case the steam and the GTUs are operated to a large extent independently. However, the heat of combustion products downstream of the GTU is utilized for heating feedwater of the STU with partial forcing out the steam regeneration. Naturally, the efficiency of this scheme is lower than that of the above options, nevertheless it exhibits a number of important advantages:

- The problem of GTU location is simplified greatly.
- There is no need in revamping the major boiler components.

The main characteristics of both units are listed in the table 3.

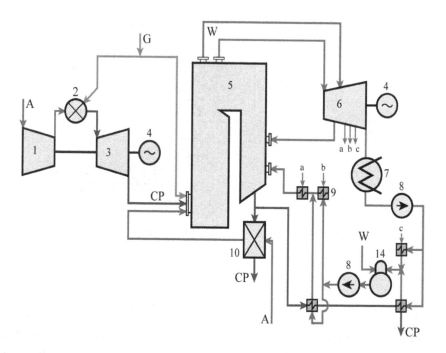

1 – compressor, 2 - combustion chamber, 3 - gas turbine, 4 - electric generator, 5 - steam boiler, 6 -steam turbine, 7 – steam condenser, 8 - feed pump, 9 - steam regeneration, 10 - air heater, 11 – deaerator. A -air, G -natural gas, CP - combustion products, GP - natural gas conversion products, W - water, steam

Fig. 4. Repowering by addition of GTU with combustion products exhaust to boiler.

Type of scheme	GTU with discharge of comb. products into the boiler furnace		GTU integrated with STU in a single steam generating circuit	
MODES	1	2	3	4
Main characteristics				
1. GTU total power, MW	48.5	48.5	48.5	48.5
2. Additional gas flow to the boiler, kg/s	14.0	12.6		
As % of the total flow	82.3	80.7		
3. Efficiency of boiler , %				
- boiler per se	88.9	87.4	94.5	94.4
- system comprising boiler+gas regenerator	88.9	93.6		
4. STU total power, MW	317.9	294.1	317.9	292.2
5. Parasitics, MW	10.1	10.1	10.4	10.4
6. Additional useful power of the unit, MW	48.5	24.7	48.2	22.5
7. Total fuel consumption by the unit, kg/s	17.0	15.59	17.99	15.86
8. Efficiency of additional electric power production, %	48.3	82.9	32.1	51.4

Mode 1 - Without gas regeneration.
Mode 2 - With gas regeneration. Maximum heat use in gas regeneration system.
Mode 3 - Separate location of GTU and STU.
Mode 4 - GTU and STU are integrated in a single steam generating circuit.
Maximum heat use in gas regeneration system.

Table 3. The main characteristics of gas turbine with discharge of combustion products into boiler and gas turbine integrated with steam turbine in a single steam-generating circuit.

2.2.5 Results of the thermodynamic analysis

The following characteristics of the various schemes were used for comparison:

1. The additional useful power output of the steam-gas unit, SGU, (Ne add, MW),
2. Efficiency of Generation of the Additional Electric Power (Eff_{add}).

$$Eff_{ad} = (N_{SGU} - N_{STU})/(\Sigma g_i \cdot Q_{iSGU} - \Sigma g_i \cdot Q_{iSTU}) \tag{2}$$

When a topping gas turbine unit is added to the STU quite natural desire is to introduce minimum changes into the flowsheet of the STU, and to retain the possibility of independent operation of the STU in case of shut-down of the GTU.

The efficiency of such repowering depends to a large extent on how successfully the following problems will be solved:

- Compatibility of the two units in flow rates of the working fluid, i.e. the possibility of passing a new volume of gases through the STU boiler so that no changes occur in the boiler working surfaces temperatures and the basic parameters of the boiler. In connection with the fact that the suggested Partial Oxidation Technology is of notably specific nature, in the calculation that will follow the "Xf" parameter will be introduced, equal to the ratio of fuel flow to the topping unit to the nominal fuel flow to the STU.

- Any one of the topping GTU causes reduction of air flow through the regenerative air heater (RAH), and by consequence additional heat losses with the flue gases. To make up for these losses some part of the flue gases upstream of the RAH may be used for heating feedwater in a special heat exchanger with corresponding forcing out the steam regeneration.
- In the topping GTU using partial oxidation the gas turbines cannot use air as coolant due to both explosion danger and cooling efficiency. Therefore, in all calculations steam extracted from the steam turbine was used for this purpose. The cooling steam flow was determined to be such as to maintain the same temperature level of the components being cooled.

When optimizing the technological scheme in certain cases it is advisable to feed additional extracted steam to the GTU; in doing so the GTU power is appreciably increased. Although the efficiency of additional electric power production in this case may be lowered the total fuel savings in the repowered unit may be larger.

In this work the amount of steam injected into the gas turbine varies as an independent variable. The diagram in Fig. 5 illustrate the effect of using various schemes of gas regenerative heating of feedwater and also different amounts of steam injected into the GTU on power and economic efficiency of different repowering topping schemes with partial oxidation based on use of the aircraft engine AL-31 GTU.

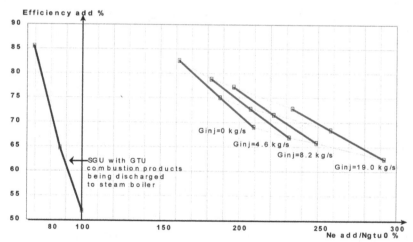

Fig. 5. Repowering by addition of AL-31 aircraft GT.

The additional power in the diagram is normalized in respect to NGTU0 -the power of the AL-31 GTU, operating in a conventional mode with one combustion chamber.

For comparison purposes the above diagrams show the indices of the scheme with the combustion products of the same aircraft GTU being discharged to the STU boiler.

2.2.6 Results of feasibility study analysis

Analysis made by OIVTRAN jointly with Mosenergoproyekt Power-Plant Design Institute reveals that the existing building of power plants may at best accommodate only one gas-turbine unit per steam-turbine plant. No acceptable technical solutions could be found involving two, to say nothing of three, gas-turbine units; in so doing, the capacity of gas turbine proper does not appear very critical (naturally, within reasonable limits).

The cost of power generation and the efficiency of capital investment are calculated with the following preconditions:

Δ_1 - the share of depreciation charges (from specific capital costs)
-    standard STU = 0.07
-    gas-turbine topping = 0.08
Δ_2- the share of maintenance repair deduction (from depreciation charges) = 0.2
Δ_3- the share of wage deduction (from specific capital cost) = 0.01
c- bank interest = 0.06
z- deductions from capital costs = 0.06
n- number of hours of operation in year = 6000
j- operational factor = 0.95

The calculations were performed for three values of the relative cost of fuel: low, medium and high. The results are given graphically in Fig. 6 (relative cost of electricity).

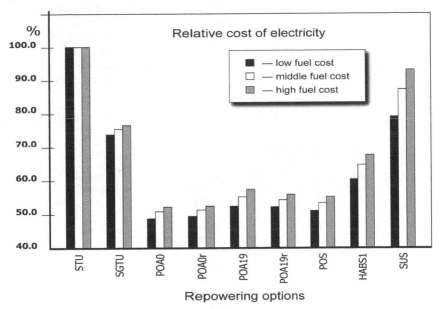

STU - conventional steam power unit, SGTU – steam-gas power unit
POA0 - topping unit, based on aircraft engine, steam injection=0; without gas regeneration
POA0r - topping unit, based on aircraft engine, steam injection=0; with gas regeneration
POA19 - topping unit, based on aircraft engine, steam injection=19 kg/c; without gas regeneration
POA19r - topping unit, based on aircraft engine, steam injection=19 kg/c; with gas regeneration
POA19 - topping unit, based on stationary gas turbine engine
HABS1 - topping unit with GTU combustion products being discharged to steam boiler
SUS - topping unit with GTU combined with STU in a single steam-generating circuit.

Fig. 6. Relative cost of electricity.

2.2.7 Conclusions

1. Repowering the existing steam-turbine plants fired with gas and fuel oil with the aid of gas-turbine toppings is much more feasible technically and economically than the closest rival option, that of construction of advanced steam-gas plants.
2. Of all of the alternative technologies for repowering the best performance is offered by facilities with partial oxidation of fuel [5]. This technology permits of:
- raising the capacity of the gas-turbine unit employed by 30-50% for heavy-duty GTU and by a factor of two or two-and-a-half for aeroderivative GTU;
- reaching the efficiency of additional generation of electricity of 60 to 80%.
- reducing to a minimum the NOx emission with the stack gas of steam power plants [6,7];
- reducing the cost of additional generated electricity two times as compared with a steam-turbine unit.

3. The creation of new highly efficient power production technologies

3.1 Environmentally friendly combined-cycle unit for cogeneration

Steam-gas plants for electric power generation have undoubted advances compared to steam-turbine plants. When efficiency of modern condenser steam-turbine plants accounts for 38-39%, the same for perspective steam-gas plants reaches 55-60%. However, when speaking about cogeneration of heat and electricity, the advantages of steam-gas plant are not that evident, because the efficiency of electric power production at heat consumption and in the nominal regime become practically equal and heat production by steam-turbine unit is much more higher than in steam-gas one [9].

For the needs of characterization of the efficiency of heat producing units, the following terms are used:

- Fuel heat utilization factor (FHUF):

$$\text{FHUF} = \frac{N_e + Q_t}{H_t} \qquad (3)$$

here FHUF - fuel heat utilization factor, N_e – useful electric power (MW), Q_t – district heating power (MW), H_t – total energy of the consumed fuel.

- Efficiency of electric power production at heat consumption regime η_e:

$$\eta_e = \frac{N_e}{H_t - \dfrac{Q_t}{\eta_k}} \qquad (4)$$

here η_e – efficiency of electric power production at heat consumption regime, η_k – efficiency of boiler of district heating facility.

The technology scheme of heat production steam-gas unit on the base of aviation AL-31 engine, produced by "Salut" Moscow Engineering Production Enterprise, was developed by JIHT RAS [10]. This unit has some advantages compared to traditional heating steam-turbines and steam-gas units. Short description and schematic diagram are given in Fig. 7.

Ambient air is compressed in the compressor НД (1) up to 6.3 kg/cm2 (abs) and fed in the mixing type air cooler (3), where the processes of saturation by steam and cooling down by evaporation take place. The use of air cooler gives the opportunity to keep the air temperature at the compressor high pressure (2) outlet at the rated level of initial gas-turbine engine (446 0C) and keep the normal mode of blade row operation.

Compressed air, natural gas and steam with temperature 285oC are fed into combustion chamber of steam-gas unit, providing the following parameters of the working fluid at the compressor common drive turbine inlet: pressure in 64 ata, temperature 1310 oC, at excess air factor α=1,12. Steam is used for the cooling of turbine group elements, thus the temperature of cooling agent can be increased over the design value for the original gas-turbine unit at keeping the temperature of blade row metal temperature below the design value.

At the outlet from the group of driving turbines, the steam-gas mixture is fed into the regenerative heat exchanger (6) where the generation and superheat of the injected steam take place together with the heating of network water for cogeneration purposes.

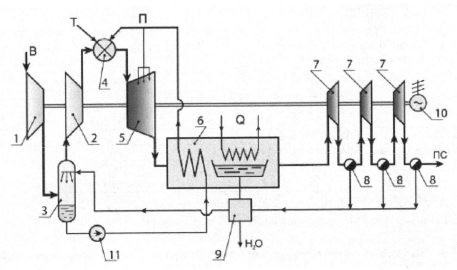

1 – low-pressure compressor, 2 – high-pressure compressor, 3 – air cooler, 4 – combustion chamber, 5 – gas turbine, 6 – heat exchanger, 7 – expander stage, 8 – drift eliminator, 9 – purifier, 10 – electric generator, 11 – feed pump, B – air, T – natural gas, П – steam, ПС – products of combustion, Q – district heating system water

Fig. 7. The scheme of steam-gas unit for co-production of heat and electricity.

The main peculiarity of the proposed scheme is higher pressure in the heat exchanger (3.05 kg/cm2 (abs)) which increases partial pressure of water vapor and gives the opportunity to condensate it at the temperature level that fits the requirements of standard district heating system for network water.

After the heat exchanger (6) steam-gas mixture is expanded up to ambient pressure in the gas expansion stage (7) with the generation of useful work and additional water due to condensation. The additional water is captured by drift eliminator (8).

Together with the main flow of condensate from the heat exchanger (6) this water is fed to the purifier (9) and after that is fed to the air cooler (3) and after that to the steam generation and steam superheat in the heat exchanger (6). The excess water condensed from the combustion products can be used for any purpose.

The steam-gas unit based on the described scheme and using the equipment of AJI-31 gas-turbine unit will have the following features (at nominal regime):

1. Useful electric power – 62.4 MW,
2. Useful thermal power – 77.8 MW,
3. Efficiency at heat consumption - 115.9%.
4. FHUF – 103.3%,
5. Air flow – 52.4 kg/s,
6. Condensate excess – 8.6 t/h.

The values of efficiency and FHUF exceed 100%, which is not a paradox and resulting from the term "the fuel lower calorific value", where the heat of condensation of steam generated at hydrogen fuel oxidation is not taken into account.

Since the products of combustion in the expansion machine are cooled below the dew point, the heat of condensation is transformed in work. The main parameters comparison for alternative heating units is given in Table 4.

Characteristics	Type of the cogeneration system			
	Steam-turbine	Combimed steam-gas plant		
	T-250/300-240	SGU 450T	SGU 230	Plant on fig.7
Net electric power, MW	231,9	433,8	213,6	62,4
Net heating power, MW	384,0	410,5	160,5	77,8
Fuel consumption, MJ/s	726,8	1033,1	460,2	135,7
FHUF	84,7	72,2	73,3	103,3
Efficiency at heat consumption	71,9	72,2	73,3	115,9

Table 4. The main parameters comparison for alternative heating units.

3.1.1 The main backgrounds of equipment selection and of the calculation of principle scheme of the unit

The main principle of the prototype selection for separate parts of the steam-gas unit and of the further calculation of thermal flow scheme is the priority of domestic aggregates and assemblies both, existing or under implementation by power engineering, i.e. the equipment without substantial difficulties concerning design and manufacture. As the main aggregate of the steam-gas unit of turbo –compressor group, the gas generator of AJI-31 gas-turbine unit, produced by NPO "Saturn" is proposed.

The preliminary analysis of the proposed scheme shows the necessity to make the following changes in the design of the original gas-turbine unit:

1. the nominal air flow should be decreased from 63.5 kg/s down to 52.4 kg/s according to the requirement of keeping the maximal mechanical stress in the blade group of compressors within the design value;

2. the low pressure compressor must be added by a group of stages to increase the total compression rate;

3. the cooling air in the cooling system of gas turbine must be replaced by steam without any change of the design. In this case the value of steam flow rate in the cooling system is defined by initial steam pressure and by hydraulic resistance of the cooling system. Numerous literature data prove that at the transition to steam cooling, the temperature of the working fluid can be increased on 90-110 °C at the same temperature of blade group metal. In the proposed scheme, the initial temperature of the working fluid is 1310 °C, being in 53 °C higher than the design value for AЛ-31 unit. In this case, the temperature of metal in the most loaded parts of steam-gas units is lower than the design value, thus increasing the unit operation reliability.

The preliminary estimations done in cooperation with the group of designers of NPO "Saturn" showed that the air-gas channel is able to let through the design flow of working fluid; the power turbine must be redesigned due to the increase of last stage blades length. The use of power turbine from another NPO "Saturn" product could be the alternative.

The designs of mixing air cooler and of heat exchangers are based on the commercial elements and have been elaborated by the specialists of NPO "Raduga".

At the calculations within the design of the expansion machine, the average aerodynamics of low-pressure cylinder stage of condensing steam turbines was taken into account with respect to the decrease of air-gas channel efficiency due to drop precipitation.

The general principle used at calculations and design is the creation of the unit consisting of restrained number of commercial parts with the opportunity of transportation by standard railroad platform. The necessity to use intensified heat exchanges, because of this, positively influenced the mass and the cost of the equipment but decreased thermal efficiency.

4. Power and Chemical Complex for co-production of electricity and synthetic liquid fuel

Power and Chemical Complex for co-production of electricity and synthetic liquid fuels (methanol, dimethyl ether, gasoline) – a promising way for saving of fuel resources.

Up to now, the progress in any branch is connected with the upgrade of one product type production technology. The thematic example for this is the progress in power production connected with the modernization of steam-gas binary cycle utilizing natural gas as a fuel. Modern steam-gas units have 55 % efficiency with the perspective of its growth up to 60%. The upgrade is taking place both due to technological scheme optimization and by the increase of actuating medium parameters at the inlet of gas turbine.

However, this way practically has reached the limit and though the further increase of working fluid parameters can increase the efficiency, the cost of produced energy most likely cannot be decreased due to investment cost growth. To our opinion the better perspective in the utilization of fossil fuels and the natural gas first of all can be obtained by the creation of power and chemical complexes with technological scheme optimized for the mass of several end products [11,12].

Certainly, it is not possible to say that in all technologies only one end product have been manufactured up to now. The production of several types of products exists but in the way of by-product production with the main mono-product being the target for the technological scheme. The basic principle for the creation of power and chemical complex technological scheme is solving several target problems by one technological process producing synergy effect by the combination of several technologies.

With this, the positive feasibility effect is obtained by the following:

- high economy efficiency due to the synergy effect induced by integration of production of several type of mass products,
- the use of the most up-to-date technologies both newly developed or taken from various branches of industry,
- multiple decrease of ecological damage to the regions of the unit sites due to decrease or in some cases total elimination of harmful emission,
- involvement, together with traditional energy sources, such primary energy resources as low- natural gas, oil gases and coal-mine methane,
- the opportunity to create feasible energy-dependent medium and small capacity plants in case of need.

The essence of the considered energy and technology complexes is the partial oxidation reaction of the initial hydrocarbons with production of energy and synthesis gas, which includes of CO and H_2. The released heat can be used for power production. The CO/H_2 ratio in the synthesis gas and the presence of H_2O, CO_2, N_2 depends on the initial hydrocarbon composition and on the technology of partial oxidation reaction.

Thermodynamic analysis and the study of kinetics of the process resulting in synthesis gas origination show that the process occurs efficiently and quickly at relatively high temperatures without catalyst. The mathematical model developed in IVTAN helps to select the optimal parameters of the process, initial component composition and determine the time process duration needed for the reaction of partial oxidation .

The first synthesis gas generator has been created in IVTAN on the base of upgraded diesel engine, where the cylinder is a unique device combining the features of compressor, high temperature chemical reactor and thermal energy to mechanical work transformer [11,13].

Fig. 8 shows the appearance of Д-245 of 80 kW diesel-based synthesis gas generator.

Fig. 8. Д-245 diesel-based synthesis gas generator.

The generator utilizes natural gas or natural gas with butane-propane thus imitating the possible compositions of oil gases. The air is used as oxidant. Excess air factor, necessary for synthesis gas production accounts for 0,4-0,6 which is absolutely inappropriate for the regular diesel operation. The transformation of diesel to synthesis gas generator demanded additional efforts, which were successful.

The synthesis gas generator in combination with catalytic reactor of methanol synthesis (Fig. 9) received the name "Sintop-300" plant. The plant capacity is ~ 800 l of methanol per day. The plant was involved in the wide range of experimental investigations.

Fig. 9. "Sintop-300" plant.

The following are the main outcomes of these investigations:

- the opportunity to organize non-catalytic conversion of natural hydrocarbon gases by partial oxidation by air at the parameters that are technically achievable,
- the possibility for efficient catalytic methanol synthesis from synthesis gas ballasted by air nitrogen,
- the possibility to create feasible energy-dependent commercial low and middle capacity units for the conversion of gaseous hydrocarbons to liquid chemical products.

It is worth mentioning that the modified diesel based synthetic gas generator is able to operate with low-pressure gases, such as off-balance or worked-out gas fields, coal bed methane, oil gases. With the use of the obtained results in 2008, the pilot plant was designed with capacity of 12000 tons of methanol per year. The main equipment was selected or partly manufactured, namely synthesis gas generator (Fig. 10), compressor, heat exchangers etc. The generator was based on 16-piston H26/26 diesel engine with net capacity 2000 KW, produced by Kolomna mechanical engineering holding.

The further development of energy and technology complexes is connected with the development of continuous-flow generator variant, allowing the use of gas-turbine and thus to increase substantially the unit capacity and start the production of the units with capacities compared with modern level of power plants. The process flow sheet of such a perspective energy and technology complex is presented in Fig.11.

Fig. 10. Synthesis gas generator.

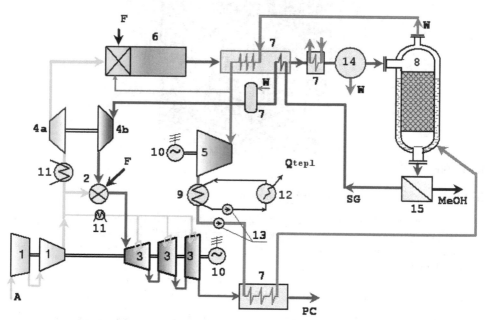

1 – compressor, 2 – the gas turbine combustor, 3 – gas turbine, 4a – air buster compressor, 4b – turbo-detander, 5 – steam turbine, 6 - partial oxidation chamber, 7- syngas cooler and waste-heat boiler, 8-synthesis CH₃OH reactor, 9 – steam condenser, 10 – electric generator, 11 – air cooler, 12 – water cooling tower, 14 – separator, 15 – product separator.

Fig. 11. Combined utilization of natural gas with the use of steam-gas turbine with the production of methanol, thermal and electric energy.

Conversion AЛ-31 aircraft engine is used as power unit. Compressed air after the compressor of the gas-turbine unit is additional compressed by compressor of the turbo-expander to 5.0 MPa and is fed to synthesis gas generator to where all the natural gas consumed by the plant is fed. The partial oxidation of natural gas is conducted in synthesis gas generator at temperatures 1200-1300 °C.

The obtained synthesis gas is cooled with the heat regeneration in power production cycle and fed in single-pass catalytic methanol synthesis reactor, where about 50-60% is converted to methanol.

The remaining gas, ballasted by nitrogen is expended in turbine of the turbo-expander and fed to the combustion chamber of gas-turbine unit.

Let us name all the advantages of the proposed scheme:

- compressed air, being the working fluid of the gas-turbine unit, is fed to the synthesis gas generator without substantial energy consumption,
- the single-pass methanol synthesis reactor with low degree of synthesis gas conversion is used, because the remaining gas is the fuel for gas-turbine unit,
- there are no energy losses at the recuperation of compressed gas energy after the methanol synthesis reactor,
- the power producing unit becomes automatically ecologically clean because the reburning of low calorific gas takes place in the gas-turbine unit combustion chamber practically without origination of toxic nitrogen oxides, both "thermal" (Zeldovich mechanism) or «prom + NOx».

The cogeneration of two products gives the opportunity to decrease considerably the cost of electricity even compared to the existing steam-gas plants. Fig. 12 shows the comparison results of relative feasibility indicators of various power plants, where the reference is the cost of electricity generated by 300 MW steam-turbine power plant.

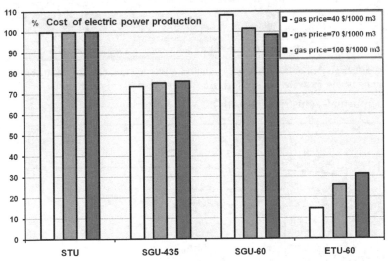

Fig. 12. The comparison results of relative feasibility indicators of various power plants.

The comparison of the cost of electricity after deduction of methanol cost of production at modern big enterprise is conducted for four alternatives at the variation of natural gas cost:

- steam-turbine power unit for supercritical steam parameters with 300 MW capacity STU,
- binary cycle steam-gas unit with capacity 435 MW (SGU 435),

- 60 MW binary cycle steam-gas unit (SGU 60),
- 62 MW(e) power and chemical complex (ETU-60).

The creation of such energy and technology complexes demands the salvation of a range of problems, where in the first place, are fine-tuning of natural gas conversion in the continuous flow reactor and the turbine operation with the use of synthesis gas ballasted by nitrogen. In order to solve these problems together with the range of other tasks, experimental-industrial unit with 1.2 MW gas turbine was created in accordance with state contract financed by the Ministry of Education and Science.

The unit includes practically all important elements of the scheme of energy and technology complex. The appearance of the unit under creation is presented in Fig.13.

1 - The synthesis gas generator by partial gas oxidation with syngas cooler, 2 – gas turbine, 3- soot filter, 4 – water heating boiler, 5 – water cooling tower, 6 – block of methanol synthesis.

Fig. 13. Experimental-industrial unit with 1.2 MW gas turbine.

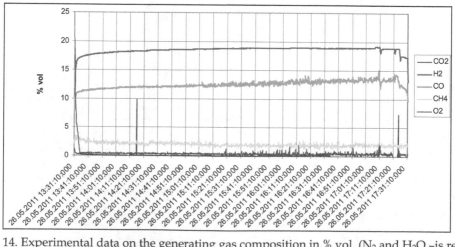

Fig. 14. Experimental data on the generating gas composition in % vol. (N_2 and H_2O –is rest).

The industrial energy and technology complex could be created, for example, on the base of 20 MW gas-turbine unit produced by "Salut" Moscow Engineering Production Enterprise. Using two of such gas-turbine units, the energy and technology complex can have electric capacity of 50 – 60 MW and produce up to 150 000 tons of methanol per year. The cost of electricity generated could be 2.5 – 3 times lower than at perspective high-capacity steam-gas unit and the exhaust gases practically free from toxic NO.

5. Total conclusions

1. The new approaches described above do not limit all the possibilities for the upgrade of the existing and for the creation of new power production technologies that increase the efficiency of natural gas utilization and decrease the harmful emission to the environment.
2. The implementation of relatively simple technical solutions in upgrade of existing power production technologies often give bigger saving rate per the investment unit, than at the creation of new traditional power production technologies.
3. The creation of energy-technology complexes of co-generation of energy, synthetic liquid fuel and other valuable accompanying products of mass demand is the perspective way for rational use of the fossil fuels (natural gas, coal, shale).

6. Acronyms and abbreviations

GTU – Gas Turbine Unit;
SGU – Combined-cycle Steam-Gas Turbine Units;
STU – Steam Turbine Unit;
Ne – electric power, MW;
H_T – energy of the fuel combusted, MW;
Q_T – thermal capacity of the boiler, MW;
η_b –boiler efficiency;
η_e – electric power production efficiency at heat combustion;
Ne add, MW- additional useful power output of the steam-gas unit after repowering;
E^{ff}_{ad} – thermal efficiency of Generation Additional Electric Power after repowering;
Ne – useful electric power (MW);
Qt – district heating power (MW);
Ht – total energy of the consumed fuel.
FHUF - fuel heat utilization factor;
N_{SGU} and N_{STU} are the useful power output of the repowered and original unit, respectively, operated at nominal rate;
$\Sigma g_i {}^* Q_{iSGU}$ and $\Sigma g_i {}^* Q_{iSTU}$ are the energy of all kinds of fuel burnt in the repowered and original units, respectively, at nominal rate;

7. References

[1] V. Bushuev, A. Troitskii "The Energy Strategy of Russia until 2020 and Real Life. What Is Next?" Thermal Engineering Vol.54, Number 1, January 2007, pp 1-7.
[2] V.M. Maslennikov, V.Ja. Shterenberg, "Method of the heat transforming into useful mechanical work in the cycle with multistage combustion", Patent of Russia 1809141 Apr., 1991/

[3] V.M. Maslennikov (USSR), R.G. Smithson (USA) et al. "Steam-and-Gas Units with Fuel Gasification within the Cycle and Ecological Problems of Power Engineering" Battele Memorial Institute, USA, 1992.

[4] V.M. Maslennikov, V.Ja. Shterenberg, "Environmentally Clean Efficient Steam - Gas Units for Repowering of Existing Power Plants", "Energeticheskoe Stroitel'stvo", Moscow, 8, 1992/

[5] V.M. Maslennikov, V.Ja. Shterenberg "Advanced gas turbine system utilizing partial oxidation technology for ecologically clean power generation", International Journal of Low-Carbon Technologies, Jan., 2011.

[6] S. Daw, K. Charkavathy, J. Pihl, J. Conkin (Oak Ridge National Laboratory) "The Theoretical Potential of Thermochemical Exhaust Heat Recuperational for Improving the Fuel Efficiency of Internal Combustion Engines", Presentation to the Chicago Section of AICHE, may 17.2011/

[7] P. Childs "Chemical looping combustion for high efficiency and carbon capture", Gas Turbine World, May-June 2011, pp 24-27.

[8] V. de Biasi "1.8:1 steam-to-fuel ratio reduces NOx output of LM 2500 to 5 ppm", Gas Turbine World, May-June 2011, pp 12-17.

[9] V. Dobrokhotov, Yu. Zeigarnik "Cogeneration: Problems and Possibilities of Realization under Present Conditions" Thermal Engineering Vol.54, Number 1, January 2007, pp 8-9.

[10] V.M. Maslennikov, Yu.A. Vyskubenko, E.A. Tsalko, V.Ya. Shterenberg. E.G. Shadek. "Heat generation in steam-gas cycle and steam-gas unit for its implementation", RF Patent №217942 F01K 26/06 of 25.04.2001.

[11] V.M. Batenin, V.M. Maslennikov, L.S. Tolchinsky "Technology of combined production of electricity and liquid synthetic fuel with the use of gas-turbine and steam-gas units", RF Patent №2250872, 27.04.2005.

[12] V.M. Batenin, V.M. Maslennikov, Yu.A. Vyskubenko "Technology for the complex use of solid fuel in power production units with cogeneration of energy and sideline products", RF Patent №2364737, 20.08.2009 г.

[13] I. I. Lishchiner, O. V. Malova, A. L. Tarasov, V. M. Maslennikov, Yu. A. Vyskubenko, L. S. Tolchinskii, Yu. L. Dolinskii "Synthesizing Methanol from Nitrogen-Ballasted Syngas" . ISSN 2070-0504 Catalysis in Industry, 2010, vol.2, No.4, pp. 368-373.

6

Use of Meso-Scale Catalysts for Bulk-Scale Processing of Natural Gas – A Case Study of Steam Reforming of Methane

Pankaj Mathure[1], Anand V. Patwardhan[2] and Ranajit K. Saha[3]
[1]Epoxy Division, Aditya Birla Chemicals Ltd.
[2]Department of Chemical Engg., Institute of Chemical Technology, Mumbai,
[3]Ex-HOD, Department of Chemical Engg., Indian Institute of Technology, Kharagpur,
[1]Thailand
[2,3]India

1. Introduction

1.1 Catalytic steam reforming

Catalytic steam reforming of hydrocarbons, alcohols and light oil fractions involves the reaction of steam with methane, ethane, natural gas, LPG, naphtha, gasoline, alcohols like methanol, ethanol, and propanol, over catalysts at elevated temperatures (473 to 1173 K) and pressures (1 to 30 atm) to produce a mixture of hydrogen, carbon monoxide and carbon dioxide (Rostrup–Neilsen, 1984) . The mixture of hydrogen and carbon monoxide is known commonly as syngas.

The technology has matured over the last 55 years. However the endothermic nature of the reactions makes the process energy–intensive. Coke formation which results in catalyst deactivation is also another major concern. Work is still in progress to use milder operating parameters, increase reactor through-put and minimise coke formation increasing catalyst life and thus reduce reformer down time.

It is reported that 90% hydrogen generated globally, is produced by the steam reforming of natural gas and light oil fractions. The major reason for this is the commercial viability of such plants by which hydrogen can be produced at $2.11/kg of H_2 (Momirlan & Veziroglu 2002, Harayanto et al. 2005). This is by far the most efficient process available in comparison to other processes like, carbon dioxide reforming, coal gasification, pyrolysis, water electrolysis and photobiochemical techniques.

1.2 Uses of hydrogen

Hydrogen has been termed as the "energy carrier of the future" (Das et al. 2001, Momirlan & Veziroglu 2002). It has the highest energy content of 120.7 kJ/gm (Harayanto et al. 2005). It burns producing no polluting emissions like SOx, NOx, CO, VOC etc. However it is available in nature only in the bound form. This makes it necessary to process the primary fuel to obtain H_2 and then use it in energy producing devices such as fuel cells or as fertiliser feedstock. Industry generates some 48 million metric tons of hydrogen globally each year from fossil fuels (Navarro et al., 2007). Nearly 50% of this hydrogen goes into making ammonia which is used in the manufacture of bulk chemicals like urea, phosphates and other fertilisers. Refineries use the second largest amount of hydrogen for chemical processes such as removing sulfur from gasoline (hydro–desulphurisation), converting heavy hydrocarbons into gasoline or diesel fuel, hydro–alkylation, hydro–cracking etc. Apart from these, hydrogen is used in Fischer–Tropsch synthesis, hydrogenation of oils and fats, pharmaceutical manufacture, cryogenic applications, in nuclear reactors, radio isotopes, mineral ore processing, reduction processes, oil processing.

1.3 Production of hydrogen

Bulk manufacturing of hydrogen is done in steam reforming units which use packed bed tubular reactors in which the catalyst (in the form of small pellets, spheres or tablets) is randomly dumped to obtain the desired products (syngas) at elevated temperature and pressure. Steam reforming system for hydrogen production is composed of a steam reformer, a shift converter, and a hydrogen purifier based on the pressure swing adsorption (PSA) or membrane separation unit. A mixture of natural gas and steam is introduced into a catalyst bed in the steam reformer, where the steam reforming reaction proceeds on nickel–based catalyst at 973 to 1073 K The reformed gas is supplied to a shift converter, where carbon monoxide is converted into carbon dioxide to produce more hydrogen by the shift reaction. The reformed gas is passed to PSA unit to separate hydrogen.

A typical steam reformer unit consists of a reformer block containing 40 to 400 tubes of height 6 to 12 m having an inner diameter of 0.7 to 0.16 m and a thickness of 0.01 to 0.02 m (Rostrup–Neilsen, 1984). The tubes are generally made of high alloy nickel chromium steel. As the process is endothermic, the required heat needs to be supplied to the reformer unit. The tubes can be heated externally or internally. In external heating, part of the feed is combusted outside the reformer tube to reach the desired reaction temperature. In internal heating part of the feed is combusted inside the tube of the reformer. This latter process is known as partial oxidation or autothermal reforming depending on the energy supplied to the process.

1.4 Catalysts used

Conventional catalysts are available in the form of cylindrical pellets, tablets and spheres and are simply dumped in the reformer tube. This is called as a fixed bed or packed bed reformer. The catalyst pellets are randomly placed in the reformer tube and do not contribute to a defined structure to the bed. There is no structure to the bed as a whole and it can be imagined as a bed containing a large number of convoluted or tortuous paths through which the feed gas flows. As the gas flows through the path, there is an increased pressure drop (relative to the use of structured catalysts). The quantity of catalyst and its composition is determined on the basis of the feed quantity and quality or composition. The role of the catalyst is to achieve maximum possible conversion of the feed to syngas, a constant pressure drop and to have sufficient mechanical and thermal strength to provide a long process cycle (Rostrup–Neilsen, 1984). Many Indian fertiliser plants use supported nickel catalysts. The support is generally silica, alumina, and/or magnesia. The properties of the catalyst are further modified by addition of components like, calcium oxide, aluminium, molybdenum, special carbides, zirconia, ceria, other alkali earth metals like potassium, etc. to make the catalyst coke resistant, and to increase its mechanical and thermal strength. Effect of promoters such as Cu, La, Mo, Ca, Ce, Y, K, Cr, Mg, Mn, Sn, V, Rh, Pd, and their combinations have been used for improving the stability of Ni catalysts supported on alumina for steam methane reforming. The use of platinum, rhodium, and palladium, along with nickel has also been reported (Harayanto et al. 2005). However, industrial catalysts are usually nickel based catalysts. The most commonly used catalysts are nickel–alumina, nickel–alumina–magnesia, nickel–magnesia, etc. The percentage of nickel varies from as low as 3 % to as high as 52% by weight. Nickel possesses hydrogenation activity but limited water gas shift activity. It is a low-cost but effective catalyst for cleavage of $O-H$, $-CH_2-$, $-C-C-$ and $-CH_3$ bonds. It is these properties that make nickel a favourable choice as a steam reforming catalyst. However, nickel alone cannot be used as a steam reforming catalyst due to its limited water gas shift activity. In order to exploit this secondary reaction, nickel or other metal catalysts are coupled (or supported on) with another metal oxide such as alumina or magnesia.

1.5 Micro and meso–scale reactors

Micro– and meso–scale reactors are a rapidly emerging technology field (Lowe, 2000, Hessel et al., 2005, Holladay et al., 2004). They differ in type / configuration and manufacturing procedures. The most widely used varieties include substrates such as silicon or silicon carbide on which channels are "etched" using well established lithography or deep reactive ion etching (DRIE) techniques. They consist of a stack of plates, onto which channels are introduced. The channels, which have width ranging from 50 to 500 μm, and depths up to 2000 μm, can be coated with active catalytic materials. This offers an advantage over traditional fixed bed reactors as higher transport rates and lower pressure drop can be achieved in a reduced volume. The large surface to volume ratio of these structures leads to effective heat dissipation making these systems suitable for highly exothermic / endothermic reactions with short contact times. There are very few papers that report the use of such micro and meso–scale reformers for the production of hydrogen from steam reforming of methane and ethanol (Pattekar & Kothare, 2004, Lowe, 2000, Hessel et al., 2005, Holladay et al., 2004).

1.6 Concept of the meso–scale, channel reactor

A novel way of developing micro or meso–scaled reactors which can be readily integrated in existing steam reformer plants for bulk manufacture of hydrogen is the use of wash–coated monoliths which can be fitted inside reformer tubes. The size of the monolith and the base material (ceramic or metal) can be selected as per the process requirements. The basic support monolith can be purchased from a number of suppliers like Corning, Degussa, and Emitec in the market and subsequently the wall coating can be modified to suit the feed stock. Monoliths or honeycomb reactors have been used as a support for catalytic converters for treatment of automobile exhaust gases and as filters for processing exhaust gas streams from industries. Monolith reactors act as multifunctional reactor systems having the advantages of low pressure, efficient contact of the reactants with (most frequently) solid catalyst and hence avoiding the problems related to partial wetting of the catalyst with liquid/gas phase; the possible choice of combining reaction and separation processes; the controlled loading of reactants, product removal, etc. Thus in monolith reactors one can combine the advantages of conventional multiphase reactors such as fixed bed, slurry and trickle bed reactors (Vergunst, 1999). This increases process efficiency and its cost effectiveness. One disadvantage of monolith reactors is the poor distribution of liquid and/or gas along the reactor's cross-section. This makes their use difficult, especially under higher flow rates.

Ceramic monoliths are usually made of cordierite which has a composition of $2MgO.5SiO_2.2Al_2O_3$. There are hardly any references (in literature) to the use of meso–scale, cordierite, monolith channels for steam reforming of methane (heat provided external to the reformer tube). It was therefore decided to use ceramic based (cordierite) monoliths having meso–scale rectangular channels as the base material for the monolithic catalyst for steam reforming of methane.

1.7 Objectives of this work

The focus of this work was on the development of novel channel monolithic reactors with meso-scale dimensions which can be coated with suitable catalytic metal oxide and used for steam reforming of methane. The objective of this article is to present the performance characteristics of such a novel reactor and to identify possible strategies for use of novel small dimension reactors for the same process. These meso-scale structures can withstand extreme pressure and temperature conditions. These structures provide the advantages of meso–scale reactors such as high surface area, low pressure drop and can be used in bulk scale manufacture of syngas and hence hydrogen. This work includes a comparison of the performance of these developed reactors with commercial catalysts (used in fixed bed reactors).

The catalyst block (named as PS–CAT) consists of a cordierite meso–scale, channeled, monolithic structure which is wash coated (a specified number of times to achieve the desired catalyst loading) with a nickel–based salt solution (2M concentration), dried, calcined, and reduced with hydrogen. This procedure of wash coating the cordierite substrate and the subsequent processing (drying and calcining of the same) is critical for the durability and quality of the catalyst (Mathure, 2008, Mathure et al., 2007a, Vergunst, 1999). The structured catalyst consists of a number of thin–walled channels which exactly fit the reformer tube inner diameter. A number of such blocks can be stacked one above the other to fill the entire length of the reformer tube. Also, the diameter of such blocks can be as per the process requirements. In this study, separate experiments were conducted with 4 channels (figure 1) as well as for 49 channels, each part of a larger monolith consisting of 100 cpsi (cells per square inch). As the reactor has straight structured channels, the resulting pressure drop is much less than that for a tubular packed bed of catalyst. This reactor concept can be used in the existing reformer infrastructure without any major change in process equipment to significantly increase hydrocarbon feed throughput. With higher throughput, the same steam reformer unit can produce larger quantities of syngas. The main advantages are increased throughput, simplicity of the arrangement, and minimum plant modification. Figure 1a shows a 4–channel PS–CAT. Figure 1b shows a schematic of the a 49–channel PS–CAT

Fig. 1a. PS–CAT (4–channel): A meso–scale, channel, structured catalyst.

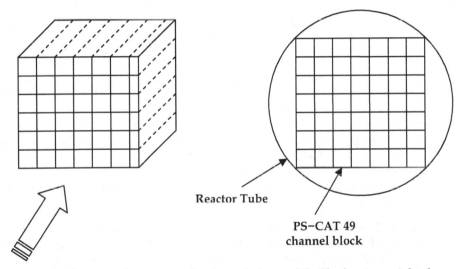

Reactor Tube

PS–CAT 49
channel block

Fig. 1b. Typical schematic of a meso–scale, channeled monolith. The figure on right shows the block placed in a reactor tube.

2. Literature review

The main focus of this section is review work on the use of monolithic, channel reactors for steam reforming of methane for production of hydrogen. Tomasic and Jovic, 2006 have published a state of the art review on the use of monolithic catalysts. According to the authors development of monolithic catalysts and/or reactors has been one of major achievements in the field of heterogeneous catalysis and chemical reaction engineering of the recent years. The work highlights the advantages of monolithic catalysts and/or reactors with respect to the conventional ones, with particular focus on the integral approach to the catalyst and reactor design. The paper is detailed and gives basic definitions and classification of monolithic catalysts, including basic features of the monoliths and factors that have proceeded to the development and application of the monolith structures. The paper also discusses the preparation of monolithic catalysts and their commercial application with particular emphasis on the less known applications, and those which are still under development. These applications include the use of monolithic catalysts for Fischer–Tropsch synthesis, selective hydrogenation, oxidation, catalytic fuel combustion units, and gas cleaning devices for industries and automobiles. The authors have concluded that futher research should be directed towards the development of new constructional designs by changing geometries to resolve the distribution problems at the entry of monolithic structures.

Hessel et al., 2005 have reviewed work on micro–structured devices (both etched channels and monoliths) developed for reforming of methane and other hydrocarbons. They have reported a number of devices which can have been used mainly for partial oxidation or autothermal reforming of methane, propane, isooctane and gasoline.

Meille, 2006 has discussed the methods used to deposit a catalyst on structured surfaces. The review outlines both physical methods such as physical vapour deposition and chemical methods (sol–gel, chemical vapour deposition, direct synthesis, etc.) The coating of catalysts based on oxide, zeolite or carbon support is detailed on various surfaces such as silicon or steel microstructured reactors, cordierite monoliths or foams, fibres, tubes, etc.

Zhou et al., 2006 have investigated alumina supported Pt group metal monolithic catalysts for selective oxidation of CO in hydrogen–rich methanol reforming gas for proton exchange membrane fuel cell (PEMFC) applications. The results show that Pt/Al_2O_3 was the most promising candidate to selectively oxidize CO from an amount of about 1 vol% to less than 100 ppm. They have investigated the effect of the O_2 to CO feed ratio, the feed concentration of CO, the presence of H_2O and/or CO_2, and the space velocity on the activity, selectivity and stability of Pt/Al_2O_3 monolithic catalysts. The Pt/Al_2O_3 catalyst was scaled up and applied in 5 kW hydrogen producing systems based on methanol steam reforming and autothermal reforming.

Giroux et al., 2005 have published a review paper for the technical and market potential of hydrogen for use in varied applications namely, stationary fuel cell power generation units, on-site generators of hydrogen for industrial uses, and hydrogen fueling stations for fuel powered automobiles. They have proposed the use of monolithic structures as alternatives to particulate catalysts for the reforming of hydrocarbons for hydrogen generation. They have further proposed that new process designs for fuel processing can take advantage of the successful experience of environmental catalysis in automotive and stationary applications using monolithic catalysts and other substrates such as heat exchangers.

Bobrova et al., 2005 have studied syngas formation by selective catalytic oxidation of liquid hydrocarbons in a short contact time adiabatic reactor. Their research involved pilot plant scale exploration of syngas formation from liquid fuels like isooctane and gasoline by selective catalytic oxidation at short contact times at nearly adiabatic reactor conditions. Monolithic catalysts with different supports (micro–channel ceramics and metallic honeycomb structure) have been used in the experiments. Their results demonstrated that over the range of the operational parameter O_2/C = 0.50–0.53 required for syngas generation, equilibrium syngas was produced over the catalysts employed, thus proving the evidence of their high activity and selectivity.

Qia et al., 2005 have studied the performance of La-Ce-Ni-O monolithic perovskite catalysts for gasoline autothermal reforming system. Autothermal reforming of gasoline or its surrogates, n–octane with or without thiophene additive, was carried out on either bulk perovskite pellet or monolithic perovskite catalyst. During the 220–h long-term test, the pellet catalyst exhibited high thermal stability and activity with hydrogen yield approaching to the theoretical maximum value and only minor amount of CH_4 slipping through. It possessed fairly good sulfur tolerance (almost immune to 5 ppmw sulfur) although it could still be seriously poisoned when subjected to high concentration of sulfur. Their raw cordierite monolith could be an effective catalyst for autothermal reforming of gasoline, exhibiting superior performance to the catalyst of 0.3 weight % Rh/CeO_2–ZrO_2/cordierite at a temperature range of 923 to 1073 K.

Pasel et al., 2004 have studied different water–gas–shift catalysts for combined autothermal reforming with water–gas–shift (WGS) reaction. The autothermal reforming of liquid hydrocarbons using different conditions have been studied. In the first step, the dry reformate from autothermal reforming and a separate stream of steam was used to conduct the WGS reaction. Strong differences concerning the catalytic activity between the three investigated commercial monolithic catalysts could be observed. The most active one showed a very promising catalytic behaviour. At a gas hourly space velocity (GHSV) of 12,250 h^{-1}, CO conversion amounted to 85% at 553 K. CO concentration in the reformate reduced from 6.1 to about 0.9 vol. %. In another step, unconverted water from autothermal reforming was fed to the reactor for the WGS reaction together with additional components of the reformate. No catalyst deactivation was observed within almost 90 h under autothermal reforming conditions generating only traces of carbon dissolved in the water.

Lindstrom et al., 2003 have carried out an experimental investigation on hydrogen generation from methanol using monolithic catalysts. They have evaluated the activity and carbon dioxide selectivity for the reforming of methanol over various binary copper–based materials, Cu/Cr, Cu/Zn and Cu/Zr. The methanol reforming was performed using steam reforming and combined reforming (a combination of steam reforming and partial oxidation). This combined reforming process was carried out at two modes of operation: near autothermal and at slightly exothermal conditions. Their results show that the choice of catalytic material has a great influence on the methanol conversion and carbon dioxide selectivity of the reforming reaction. The zinc–containing catalyst showed the highest activity for the steam reforming process, whereas the copper/chromium catalyst had the highest activity for the combined reforming process. The copper/zirconium catalyst had the highest CO_2 selectivity for all the investigated process alternatives.

Apart from these there are a few patents which refer to the use of ceria–coated zirconia monoliths for catalytic partial oxidation process of hydrocarbons for syngas production and the use of platinum and palladium monoliths for autothermal reforming of hydrocarbons (Jiang et al, 2006, Hwang et al., 1985).

Vergunst, 1999 has published a detailed thesis on the preparation aspects of carbon–coated monolithic catalysts and three phase hydrogenation of cinnamaldehyde using the same [3]. The author has covered preparation of monolithic catalysts, testing of monolithic catalysts in the (selective) hydrogenation of benzaldehyde, cinnamaldehyde, methylstyrene, benzene, and γ–butyrolactone, testing of monolithic catalysts in solid acid catalyzed reactions (acylation and esterification) and in the selective oxidation of cyclohexanol into adipic acid, development of internally finned monolithic structures, hydrodynamics, and scale up. The work is extremely important for any work related to monolithic catalysts. However the manufacture of these catalysts and their subsequent application is totally different from the current application (steam reforming) being discussed.

As can be observed from the above work, wash-coated monoliths with meso or micro-scaled channel sizes are a promising substitute for conventional large and small scale fixed bed reactors. Once the coating procedure and coating material is finalized, these monoliths provide cost-effective reactor geometries to retrofit existing reformer units. To the best of our knowledge, there are hardly any references (in literature) to the use of meso-scale, ceramic, monolith channels for steam reforming of methane (heat provided external to the reformer tube). It was therefore decided to use ceramic based (cordierite) monoliths having meso-scale rectangular channels as the base material for the monolithic catalyst for steam reforming of methane. As nickel possesses hydrogenation activity and is a low-cost but effective catalyst for cleavage of $O-H$, $-CH_2-$, $-C-C-$ and $-CH_3$ bonds, it was decided to exploit these properties and use nickel nitrate solution for the coating of the monoliths.

3. Development of the monolith structures

3.1 Coating of the monolith structures

The catalyst block (named as PS-CAT) developed in the course of our work consisted of a cordierite substrate with a square-channel, monolithic structure which was wash-coated with a nickel-based salt solution (2M concentration), dried at 393 K, calcined at 673 K, and reduced with hydrogen (Mathure, 2008). The catalyst block is shown in figure 1. The structured catalyst consisted of a number of thin-walled channels which exactly fitted the experimental reformer tube inner diameter (Mathure et al. 2007b). The average wall thickness was found to be about 270 μm. The channel dimensions were determined to be 1500 μm × 1500 μm (hence the term meso-scale). In this study, experiments were conducted with a block having 4 channels cut out from a larger monolith consisting of 100 cpsi (channels per square inch). Figure 1 shows a 4-channel PS-CAT. The performance of the developed catalyst was tested extensively using a multi-feed, multi-scale, fully instrumented experimental rig fabricated by M/s. Texol Engineering Private Limited, Pune, Maharashtra, India & M/s Dampf Kolben , Pune, India. The details of experimental rig and procedure have been reported elsewhere (Mathure et al., 2007b, Mathure, 2008). Figure 2 from shows the actual photograph of the experimental rig and its components such as the mass flow meter, HPLC pumps, reactors and Reactor furnace assembly. Figure 3 shows the schematic arrangement of the experimental rig. PS-CAT contained a nickel oxide loading of around 8.5% (weight% of the coated block). The elemental nickel loading was around 6.7% (weight% of the coated block) as determined by EDS –X-ray analysis.

Each block was weighed before being dipped into the coating solution, and after the calcination, to find out the amount of catalyst (nickel oxide) loading on the block. This weight was taken as W for calculation of W/F_{CH4} during the steam reforming experiment. The blocks were activated using a hydrogen stream of flow rate 40 N ml/minute in the steam reformer unit just before use in the experiment.

Fig. 2. Experimental rig and its components used for steam reforming of methane.

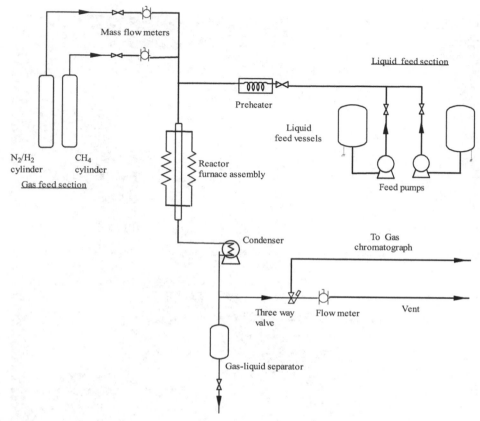

Fig. 3. Schematic of the experimental rig used for steam reforming of methane and ethanol.

3.2 Characterisation of the structures

The characterisation was done by EDS–X-ray analysis, BET surface area measurement and XRD analysis. A detailed discussion of the characterization can be found the work of Mathure, 2008.

3.2.1 X-ray diffraction study

The crystal phases in the uncoated cordierite substrates and the coated PS–CAT were identified by a X-ray powder diffraction study. XRD spectra were obtained with a Rigaku–Miniflex diffractometer using monochromatic Cu–Kα radiation. The 2θ scanning was performed from 10 to 80° at a rate of 2°/min. Figure 4 shows the XRD patterns for the uncoated cordierite substrate and Figure 5 shows the XRD pattern for the coated PS–CAT. The only active ingredient in the coated catalyst was the nickel oxide species. An observation of both XRD patterns shows that the extra peaks present in Figure 5 are of nickel oxide as verified using JCPDS database. These peaks are not seen in the uncoated raw substrate.

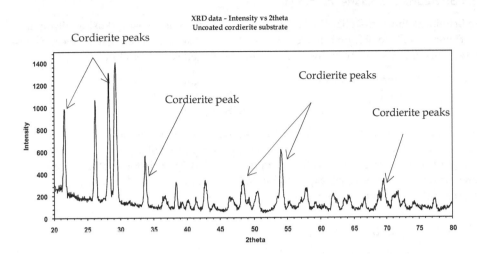

Fig. 4. X-ray diffraction spectra of the raw uncoated cordierite substrate.

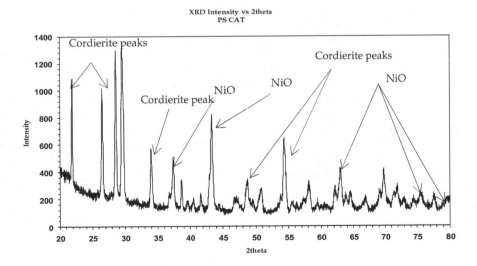

Fig. 5. X-ray diffraction spectra of the coated catalyst block PS–CAT.

3.2.2 Energy dispersive spectroscopic X–ray analysis (EDS–X-ray)

The metal loading was estimated by EDS– X–Ray unit (Link ISIS 300, Japan). Nickel oxide which acts as the active ingredient content was found to be 8.5% (weight % of the coated block). The elemental nickel loading was found to be around 6.7% (weight % of the coated block).

3.2.3 BET surface area measurement

The Brunauer-Emmett-Teller (BET) specific surface area was measured using nitrogen as an adsorbent in a Quantachrome make AUTOSORB 1C machine. The samples were degassed for 12 hours at 573 K before analysis. The values obtained were as follows;

Uncoated (raw) cordierite substrate	: 3.14 m²/gm
Coated substrate, PS–CAT	: 1.59 m²/gm
Used (post–experimentation) PS–CAT	: 1.18 m²/gm

The extremely low values have been observed due to the use of cordierite ceramic blocks which have very low specific surface area in the first place. Subsequent and prolonged dip coating further decreases this value. It must be noted here that the blocks were directly coated with the metal salt solution. No binder or support material such as γ-Al$_2$O$_3$ sol was used to enhance the surface area.

4. Experimental study

The performance of the developed catalyst was tested extensively using the experimental set up described detail in Mathure, 2008 & Mathure et al., 2007a. The testing was carried out by varying the process parameters such as time–on–stream analysis (100 hours on stream), temperature (673 to 1048 K), space time (W/F$_{CH4}$ = 10 to 55.0 gm-cat min/mol), and the number of channels (4 channels and 49 channels in parallel). However as the number of monoliths available to us were limited, the bulk of the experiments were performed using the 4 channel PS–CAT. Time–on–stream studies were carried out for both the blocks. The length of each block (both 4 channels and 49 channels) was 0.025 m. In case of experiments with the 4 channel PS–CAT, the tubular reactor housing with inner diameter 6 mm was used. Five blocks each having 4 channels were placed end to end within the tubular reactor housing such that the centre of the catalyst bed corresponded to the central heating zone of the furnace. Rest of the tubular reactor housing (both above and below the catalyst blocks) was filled up with inert particles. As 5 blocks were used in series, the total length of the meso–scale channel reactor was 0.125 m. In case of experiments with the 49 channel PS–CAT, the tubular reactor housing with inner diameter 22 mm was used. Two blocks each having 49 channels were placed end to end within the tubular reactor housing such the centre of the catalyst bed corresponded to the central heating zone of the furnace. Rest of the tubular reactor housing (both above and below the catalyst blocks) was filled up with inert particles. As 2 blocks were used in series, the total length of the meso scale channel reactor was 0.05 m. The reactor was mounted in the furnace assembly. Fresh catalyst was used for each experimental run.

The catalyst was activated in a hydrogen flow of 40 N ml/min with a steady increase of temperature. The catalyst was maintained at the desired reaction temperature for two hours. Methane gas (make: BOC, XL grade, purity: 99.9 %, specific gravity: 0.553, calorific value: 55530 kJ/kg) was fed to the reactor at the desired flow rate using a Bronkhorst make mass flow meter. The preheater temperature was maintained at 443 K to ensure complete vaporisation of the feed water even at high liquid flow rate.

A HPLC (high pressure liquid chromatography Make: SSI, USA) pump was used to feed water to the reactor at the desired flow rate (between 0 to 5 ml/min). The product stream was condensed using the double pipe condenser at the outlet of the reactor. The liquid product (mostly water) was collected in a separator while the outlet gas stream flow rate was measured using a wet gas flow meter.

The gas was periodically sent to the GC (gas chromatograph) for analysis using a three-way valve. Gas analysis was done using a thermal conductivity detector (TCD) with Spherocarb column (1/8″ diameter and length 8′) to find the composition of the outlet gas stream. The product gas was found to contain methane, carbon monoxide, carbon dioxide and hydrogen. No other hydrocarbon species was detected. The results of the gas analysis of the product gases indicated whether steady state conditions had been achieved. The measurements for the study were done once similar gas compositions were obtained after the start of the experiment. As this is a fast reaction in general steady state conditions were observed within 15 to 20 minutes from the start of the experiment. The pressure for all experimental runs was maintained at 1 atmosphere.

The liquid product was weighed and analysed periodically in the GC using the TCD in a Porapak Q column (1/8″ diameter and length 6′). The liquid product was found to be only water. The reaction was stopped after 3 to 6 hours of steady state operation (except for time–on–stream studies which were conducted for 100 hours of operation).

Methane conversion was calculated using equation 1.

$$\frac{mols_{CH_4}in - mols_{CH_4}out}{mols_{CH_4}in} \times 100 = X_{CH_4} \tag{1}$$

Reproducibility of the experiments was verified by conducting repeat runs for a few random reaction conditions. The results showed a maximum error of ±4.0%.

5. Results and discussion

5.1 Effect of temperature

Steam reforming of methane was carried using PS–CAT in a temperature range of 673 K to 1048 K keeping the molar ratio of feed to steam (MR) and the W/F_{CH4} constant. Figure 6 shows the effect of temperature on the conversion of methane. As temperature increases the conversion of methane gradually increases from 30% to nearly 100%. The catalyst exhibited a methane conversion of more than 90% for a temperature above 973 K. The catalyst showed nearly complete conversion (above 98 %) of methane at a temperature above 1048 K at a W/F_{CH4} close to 55.0 g–cat min/mol and a molar ratio ($CH_4:H_2O$) close to 3.0. With increase in temperature the carbon monoxide content in the product gas stream increase from as low as 0.21% to 22.43% (v/v). The carbon dioxide content decreases with increase in temperature. Thus the catalyst promotes the endothermic steam reforming reaction but does not promote the exothermic water–gas–shift reaction which is favoured at low temperature. This result is in accordance with the presence of nickel as the active ingredient of the catalyst which promotes bond cleavage in the steam reforming reaction but has limited water–gas–shift activity (Rostrup–Neilsen, 1984, Vaidya & Rodrigues, 2006).

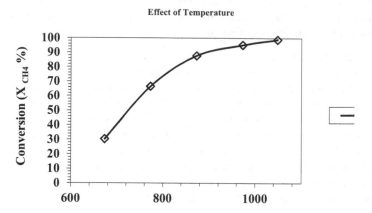

Fig. 6. Effect of temperature on methane conversion. Other conditions: PS–CAT 4 channel block, 5 blocks in series, W/F$_{CH4}$ = 56 gm–cat min/mol, steam to methane molar ratio in feed = 3:1.

5.2 Time–on–stream study

The most important feature of industrial catalysts is their ability to perform steam reforming of methane for hundreds of hours of continuous operation without significant loss of activity (Rostrup–Neilsen, 1984). The study to test the performance of the catalyst for long duration of continuous operation is called as time–on–stream study.

PS-CAT was tested for steam reforming of methane at a steam to methane molar ratio of 3:1 at atmospheric pressure and temperature of 1048 K for a over 100 hours on stream. Both the 4 channel PS-CAT as well as the 49 channel PS-CAT were tested. The resulting exit gas was analysed and the results of the same for four channels is shown in figure 7 and figure 8. As can be seen a methane conversion of more than 98 % is obtained over the entire period (more than 100 hours) of operation. The product gas composition also shows steady formation of all species. In case of the 49 channel PS-CAT the temperature of operation was 1113 K. The higher temperature was necessary as due to increase in the number of channels, the time required to acquire steady state conditions increased by 60 minutes. The reason for this is the low thermal conductivity of the cordierite substrate (2.76 W/m K). Thus the 49 channel block required more time to "soak– up" heat and attain a uniform temperature. Once a uniform temperature was attained, the performance of this block was also very good and exhibited a steady methane conversion of more than 97%. Therefore the catalyst block can readily be used in industrial reformers simply by using a block of higher diameter to fit the inside of a reformer tube with minimum modification to the existing infrastructure of the plant. In both cases the amount of CO produced was around 20 % (v/v of the dry product gas). This shows that the catalyst exhibits limited water gas shift capability. This aspect has been included in the scope of future work wherein the catalyst composition can be varied to enhance water gas shift reaction also. The amount of hydrogen produced was nearly 70 % (v/v of the dry product gas) in both cases. Thus PS-CAT is a very promising substitute for packed bed reactors for production of syngas and hydrogen which can be used for manufacture of fertilisers.

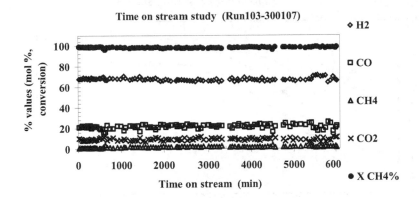

Fig. 7. Product gas composition of the exit gases. Other conditions PS–CAT 4 channels, 5 blocks in series, feed to steam molar ratio of 1:3, W/F = 56 gm–cat min/mol, pressure = 1 atm and temperature 1048 K, X_{CH4} = 98.27%, Yield 3.4 mol of H_2/mol of CH_4 reacted.

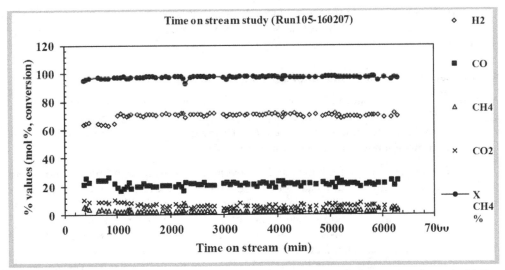

Fig. 8. Product gas composition of the exit gases. Other conditions PS–CAT 49 channels (7×7) channels, 2 blocks in series, feed to steam molar ratio of 1:3, W/F = 56 gm–cat min/mol, pressure = 1 atm, T =1113 K, X_{CH4} = 98.27%, Yield 3.14 mol of H_2/mol of CH_4 reacted.

5.3 Effect of space time

The effect of space time was studied by varying the W/F_{CH4} ratio from 10 to 55.0 gm–cat min/mol for three temperatures namely 873, 973 and 1048 K at a molar ratio of steam to methane of 3.0. Figure 9 shows the effect of space time on methane conversion. The

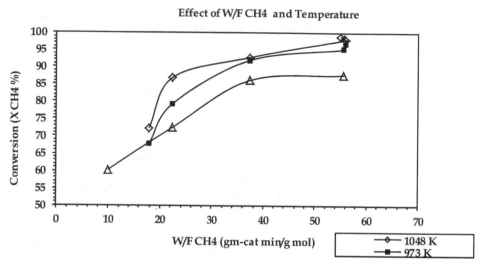

Fig. 9. Effect of space time on methane conversion. Other conditions: PS–CAT 4 channel block, 5 blocks in series, steam to hydrocarbon molar ratio in feed = 3.0.

conversion of methane increases with increase in W/F_{CH4} ratio. The catalyst exhibits a conversion of more than 90% for a W/F_{CH4} ratio above 37.3 gm–cat min/mol for temperature above 923 K.

5.4 Comparison of performance with a commercial catalyst

As mentioned in the literature review section, to the best of our knowledge there are hardly any references (in literature) to the use of meso–scale, ceramic, monolith channels specifically for steam reforming of methane (heat provided external to the reformer tube). Thus direct comparison with other data was not possible.

In order to compare the performance of the developed catalyst, experimental runs were conducted using a commercial nickel containing catalyst in a tubular fixed bed reactor of inner diameter 6 mm and 3 mm. In case of the 3 mm and 6 mm reactors the experiment was conducted maintaining the ratio of the length of the catalyst bed to the diameter of the reactor to enable a common basis of comparison between the two cases (Hoang and Chan, 2006). This ratio was maintained at 10. Table 1 gives the details of the process parameters used to compare the performance of PS–CAT with a commercial catalyst. The table also gives the composition of the catalyst determined using EDS–X-ray analysis. Swamy, 2007 has reported a detailed kinetic study of steam reforming of methane using the same catalyst. The commercial catalyst was crushed, sieved to obtain the required size of particles (approx 1.0 mm). The required weight of catalyst (0.1 gm) was filled in the tubular packed bed reactor of diameter 3 mm or 6 mm such that the centre of the catalyst bed corresponded to the central heating zone of the furnace. The remainder of the reactor was filled with inert ceramic material. The reactor was mounted in the furnace assembly. The catalyst was activated in a hydrogen flow of 40 N ml/min with a steady increase of temperature. The

Composition of active components		Structure	Weight	Residence Time	Molar Ratio	Temp.	Conversion of Methane
			gm	gm cat min / mol	Steam to Methane	K	(%)
Commercial Catalyst							
compound	wt– %	Particles of average size of 1 mm. Packed bed reactor used	0.1	56.00	3	1048	**79.73**
Ni	25						
Al₂O₃	15						
MgO	55						
PS-CAT							
compound	wt– %	Meso Scale Channel dimension 1.5 x 1.5 mm 100 cpsi 4 channels used, 5 blocks of 25 mm length used in series	0.1 (coating only)	56.00	3	1048	**98.27**
NiO	8.5						

Table 1. Process parameters used to compare the performance of PS–CAT with a commercial catalyst.

catalyst was maintained at the desired reaction temperature for two hours or till steady state conditions were achieved which can be determined from the temperature of the bed and the reactor skin temperature (outer wall of the reactor). At identical conditions (feed to steam molar ratio of 1:3 at a W/F = 56 gm–cat min/mol , pressure = 1 atm and temperature 1048 K), when a commercial nickel-based steam reforming catalyst was tested for steam reforming of methane in a 6 mm packed bed reactor the corresponding conversion obtained was 79.7 %. The conversion of methane using PS–CAT at identical conditions was found to be 98.27 % exhibiting a massive increase in the conversion of methane. Figure 10 shows the effect of temperature on methane conversion while comparing the performance of PS–CAT with commercial catalyst at identical process conditions. Figure 11 compares the performance of PS–CAT with commercial catalyst exhibiting the effect of W/F_{CH4} on methane conversion. In both cases, PS–CAT outperforms the commercial catalyst.

The reactor exhibited a minimum increase of 10% and a maximum increase of 22% in the conversion of methane that directly translates into increased production of syngas and hydrogen with existing plant infrastructure. Thus PS–CAT is a very promising substitute for packed bed reactors for production of syngas and hydrogen which can be used for manufacture of fertilisers. The better contact between the reactants and the active catalytic ingredient in case of the use of the structured catalyst shows a substantial increase in the performance in comparison to a fixed bed catalyst consisting of randomly packed catalyst. In the case of the 3 mm and 6 mm diameter fixed bed reactors, the conversion of methane in the 3 mm reactor was higher than the 6 mm reactor. This is in agreement with principles of use micro or meso–scale devices (Lowe, 2000). Decreasing the linear dimensions, for a given

Comparison of performance with commercial catalyst

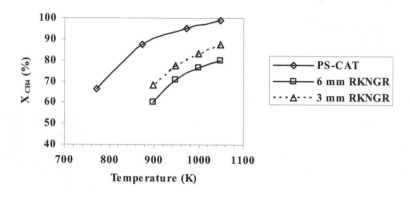

Fig. 10. Comparison of performance of PS–CAT with commercial catalyst, effect of temperature on methane conversion. Other conditions: PS–CAT 4 channel block, 5 blocks in series, W/F_{CH4} = 56 gm-cat min/mol, steam to hydrocarbon molar ratio in feed = 3.0.

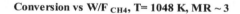

Conversion vs W/F $_{CH4}$, T= 1048 K, MR ~ 3

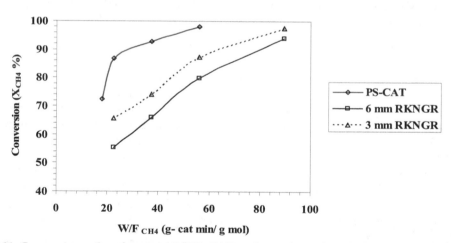

Fig. 11. Comparison of performance of PS–CAT with commercial catalyst, effect of W/F_{CH4} on methane conversion. Other conditions: PS–CAT 4 channel block, 5 blocks in series, T= 1048 K, steam to hydrocarbon molar ratio in feed = 3.0.

difference in a physical property (like temperature or concentration) increases the respective gradient. Consequently the driving forces for heat transfer, mass transport or diffusion flux per unit volume or unit area increase.

Literature reports the effect of changing diameter of a membrane reactor system on methane conversion for the partial oxidation process (Hoang and Chan, 2006). The authors have reported an increase in methane conversion with decrease in reactor diameter (at the same length to diameter ratio). Even though the system reported is different from this study the reasoning for the increase in methane conversion for smaller reactor diameters has a similar basis and can explain the increase in conversion for the 3 mm reactor.

6. Conclusions

A novel, meso–scale, structured, channel reactor was developed and successfully tested for steam reforming of methane. The structured catalyst called PS–CAT is a cordierite based monolithic structure coated with nickel oxide. The coating procedure was finalised on the basis of a set trials. The characterisation was done by EDS–X-ray analysis, BET surface area measurement and XRD analysis. The testing was carried out by varying the process parameters such as time–on–stream analysis (100 hours on stream), temperature (673 to 1048 K, space time (W/F_{CH4} = 10 to 55.0 gm–cat min/mol), and the number of channels (4 channels and 49 channels *in parallel*). PS–CAT was tested for steam reforming of methane at a steam to methane molar ratio of 3:1 at atmospheric pressure and temperature of 1048 K for over 100 hours of operation.

The conversion of methane observed with the said conditions was above 97 %. The catalyst exhibited good conversion (above 90%) of methane for a molar ratio of steam to methane from 2.5 to 7. The catalyst showed a conversion of more than 90% for a W/F_{CH4} ratio above 37.3 gm–cat min/mol for temperature above 923 K. Maximum hydrogen yield of 3.8 mol of H_2/mol of CH_4 reacted was observed at a W/F_{CH4} of 55.1 gm–cat min/mol at a temperature of 1048 K and molar ratio of steam to methane of 3.0.

Finally the performance of this developed catalyst was compared with the performance of commercial nickel–based catalyst in small diameter (3 mm and 6 mm) fixed bed reactors. The catalyst exhibited a minimum increase of 10% and a maximum increase of 22% in the conversion of methane that directly translates into increased production of syngas and hydrogen with existing plant infrastructure.

The main advantages of use of PS–CAT are increased throughput, simplicity of the arrangement, and minimum plant modification. Thus PS–CAT is a very promising substitute for packed bed reactors for production of syngas and hydrogen which can be used for manufacture of fertilisers. This case study also emphatically demonstrates the application of meso-scale catalytic structures for bulk scale production of commodity gases such as syngas

7. Nomenclature

F_{CH4}	=	molar flow rate of methane, mol/min
F_{H2O}	=	molar flow rate of water, mol/min
Q_{CH4}	=	volumetric flow rate of methane, N ml/min
Q_{H2O}	=	volumetric flow rate of water, ml/min
T	=	temperature, K
v/v	=	composition of gas species expressed as volume percent of the total product gas %
W	=	weight of catalyst, gm
W/F	=	ratio of weight of catalyst to molar flow rate of feed (space time), gm–cat min/mol
X_{CH4}	=	fractional conversion of methane

8. Abbreviations

BET–SSA	=	Brunauer–Emmett–Teller specific surface area method
cpsi	=	cells per square inch
EDS X-ray	=	energy dispersive spectroscopic X–ray analysis
GC–TCD	=	thermal conductivity detector of gas chromatograph
MR	=	molar ratio of steam to feed hydrocarbon
XRD	=	X–ray powder diffraction
WGS	=	water–gas–shift
DRIE	=	deep reactive ion etching
LTS	=	low temperature shift
PSA	=	pressure swing adsorption

9. Acknowledgments

I wish to acknowledge the Department of Fertilizers, Ministry of Chemicals and Fertilizers, Government of India, for providing the financial assistance for the execution of the project titled, *"Studies on reforming of methane to synthesis gas using micro–reactors for production of hydrogen (MPH)"*, sanctioned *vide* letter - 15011/1/2004–FP, dated 05/03/2004.

I wish to thank the Council of Scientific and Industrial Research (CSIR) for the award of Senior Research Fellowship (Individual) sanctioned *vide* letter 9/81(624)/07 EMRI dated 09/03/2007 from April 2007 onwards.

I would like to thank my co-authors & my ex-colleague Prof. Anand V. Patwardhan, Prof. Ranajit K. Saha & Shouvik Ganguly for their invaluable insights and unwavering support.

I would like thank my sister & my brother-in-law Tejas & Sameer Oundhakar for providing financial assistance in the publication of this book chapter.

10. References

Bobrova, L., Zolotarskii, I., Sadykov, V., Pavlova, S., Snegurenko, O., Tikhov, S., Korotkich, V., Kuznetsova, T., Sobyanin, V. and Parmon, V., Syngas formation by selective catalytic oxidation of liquid hydrocarbons in a short contact time adiabatic reactor, *Chemical Engineering Journal*, (2005), *107*, 171–179.

Das, D., Veziroglu, T. N., Hydrogen production by biological processes: A survey of literature, *International Journal of Hydrogen Energy* (2001), *26*, 13–28.

Giroux, T., Hwang, S., Liu, Y., Ruettinger, W. and Shore, L., Monolithic structures as alternatives to particulate catalysts for the reforming of hydrocarbons for hydrogen generation, *Applied Catalysis B: Environmental*, (2005), *56*, 95–110.

Harayanto, A., Fernando, S., Murali, N., Adhikari, S., Current status of hydrogen production techniques by steam reforming of ethanol: A review, *Energy and Fuels* (2005), *19*, 2098–2106.

Hessel, V., Lowe, H., Muller, A. and Kolb, G., *Chemical Micro Process Engineering – Processing and Plants*, Wiley–VCH, Germany (2005).

Hoang D.L., Chan S.H., Effect of reactor dimensions on the performance of an O_2 pump integrated partial oxidation reformer–a modeling approach, *International Journal of Hydrogen Energy*, (2006), *31*, 1–12.

Holladay, J.D., Wang, Y., Jones, E., Review of developments in portable hydrogen using microreactor technology *Chemical Reviews* (2004), *104* , 4767–4790.

Hwang, H., Heck, R.M., Yarrington, R. M., Fuel cell electric power production *United States Patent 4,522,894*, (1985)

Jiang, W., Park, S., Tomczak, M. S., Acharya, D. R., Tamhankar, S. S., Ramachandran R., Monolith based catalytic partial oxidation process for syngas production *United States Patent 7,090,826*, (2006)

Lindström, B., Agrell, J., Pettersson, L.J., Combined methanol reforming for hydrogen generation over monolithic catalysts, *Chemical Engineering Journal*, (2003), *93* ,91–101

Lowe H. V., *Microreactors New technology for modern chemistry*, Wiley–VCH, Germany (2000).

Lowe, H. V., *Microreactors New technology for modern chemistry*, Wiley–VCH, Verlag GmbH, Weinheim, (2000).

Mathure P.V., Ganguly S., Patwardhan A.V. and Saha R.K., Steam reforming of ethanol using a commercial Nickel-based catalyst, *Industrial and Engineering Chemistry Research* (2007b), *46*, 8471-8479.

Mathure P.V., Swamy B., Patwardhan A.V., Saha R.K., Ganguly S., A structured catalyst for steam reforming of methane for production of syngas *Application filed in the name of I.I.T., Kharagpur –Indian Patent Application No. 1515/KOL/2007* dated 02/11/2007a.

Mathure, P.V., Studies on steam reforming for the production of hydrogen, Ph.D. Thesis, I.I.T., Kharagpur, India (2008).

Meille, V., Review on methods to deposit catalysts on structured surfaces, *Applied Catalysis A: General*, (2006), *315*, 1–17.

Momirlan, M., Veziroglu, T.N., Current status of hydrogen energy, *Renewable and Sustainable Energy Reviews* (2002), *6*, 141–179.

Navarro, R. M., Pena, M. A., Fierro, J. L. G., Hydrogen production reactions from carbon feedstocks: fossil fuels and biomass, *Chemical Reviews* (2007), 107, 3952–3991.

Pasel, J., Cremer, P., Wegner, B., Peters, R. and Stolten, D., Combination of autothermal reforming with water–gas–shift reaction – small scale testing of different water–gas–shift catalysts, *Journal of Power Sources*, (2004), *126*, 112–118.

Pattekar, A.V., Kothare, M.V., A microreactor for hydrogen production in micro fuels cell applications, *Journal of Microelectromechanical Systems* (2004), *13*, 7–18.

Qia, A., Wang, S., Fub, G., Nib C. and Wub, D., La–Ce–Ni–O monolithic perovskite catalysts potential for gasoline autothermal reforming system", *Applied Catalysis A: General*, (2005), *281*,233–246.

Rostrup–Neilsen, J.R., Catalytic Steam Reforming in *Catalysis, Science and Technology, vol.5,* ed. Anderson, J. R. and Boudart, M., Springer–Verlag, Berlin (1984).

Swamy, B., *Studies on methane steam reforming for production of hydrogen*, Master's Thesis, I.I.T., Kharagpur, India (2007).

Tomasic, V., Jovic, F., State-of-the-art in the monolithic catalysts/reactors, *Applied Catalysis A: General* (2006), *311*, 112–121.

Vaidya, P.D., Rodrigues, A.E., Insight into steam reforming of ethanol to produce hydrogen for fuel cells, *Chemical Engineering Journal* (2006), *117*, 39–49.

Vergunst, T., *Carbon coated monolithic catalysts preparation aspects and three phase hydrogenation of cinnamaldehyde*, PhD Thesis, Technical University, Delft, The Nethelands (1999).

Zhou, S., Yuan, Z. and Wang, S., Selective CO oxidation with real methanol reformate over monolithic Pt group catalysts: PEMFC applications, *International Journal of Hydrogen Energy*, (2006), *31*, 924 – 933.

Part 3

Natural Gas Combustion

7

The Influence of Modified Atmosphere on Natural Gas Combustion

Małgorzata Wilk and Aneta Magdziarz
AGH University of Science and Technology, Krakow,
Poland

1. Introduction

Coal technology, dominant in Poland, ensures efficient production of electricity and heat. However, a large exploitation of existing reserves and the growing demand for electricity causes interest in other available fuels, primarily natural gas. Environmentally friendly modern technologies are constantly looking for the possibility of using energy sources other than coal. Research is carried out on various innovative technologies such as CO_2 capture and storage or unconventional combustion. Combustion in the oxygen-enriched gas mixtures at the first was conducted in the steel industry and metallurgy, which require very high temperatures to heat the metal and pig iron. Such a process can be classified as the future technology known as the clean combustion technologies.

The use of natural gas in the metallurgical processes requires the precise technological regime, therefore, it is difficult to make changes in the process (Amann, Kanniche & Bouallou, 2009; Choi & Katsuki, 2001; Flamme, 2001). However, appropriately organized gas combustion process can lead to a reduction in CO_2 emissions, which is required by applicable laws imposed by the European Union. The European Union environmental priorities promote the development of the new technologies of energy production from the conventional sources (coal and natural gas), e.g. oxy-combustion (Andersson & Johnsson, 2007; Czakiert et al., 2006; Davidson, 2007; H.K. Kim, Kim, Lee & Ahn, 2007; Kotowicz, 2007; Lampert & Ziębik, 2007; Li, Yan & Yan, 2009; Muskał et al., 2008; Seepana & Jayanti, 2009; Simpson & Simon, 2007; Szlęk et al., 2009; Tan et al., 2002). The oxy-fuel combustion process is conducted in an atmosphere enriched in oxygen which means that the reactor is supplied by the combustion gas mixture, in which the oxygen concentration is higher than the concentration of oxygen in the air. The study of the modified air combustion appears in scientific literature. There are various modified oxidizing atmospheres like O_2/N_2, $O_2/N_2/CO_2$, O_2/CO_2. It should be emphasized that the majority of results of the oxy-combustion process refer to coal combustion, because of large amount of deposited coal in Poland (Buhre et al., 2005; Chen, Liu & Huang, 2007; Croiset, Thambimuthu & Palmer, 2000; Czakiert, Nowak & Bis, 2008; Kim et al., 2007; Normann et al., 2008). The research effort is focused on the conventional coal, fluidised bed combustion and co-generation solutions with the coal gasification (Lampert & Ziębik, 2007; Tan et al., 2002). Oxy-combustion is the subject of research in many international research institutions (Chalmers University of Technology, Sweden, University of Leeds UK, CANMET Energy Technology Centre, Canada, Chicago USA Research Centre, Tokyo Institute of Technology, Japan, University of Newcastle Australia). In Poland, the research of oxy-combustion concerns mainly coal (Czestochowa University of Technology), but the natural gas research is also carried out in the field of high-temperature combustion gas HTAC technique (Normann, 2008; Seepana & Jayanti, 2009). Fuel combustion processes are the main source of environmental pollution. In many branches of industry, the main fuel is natural gas, because of the possibility of obtaining a high temperature process e.g. in the glass industry and in the manufacture of cement, the process of oxy-combustion can be applied, or air combustion enriched with oxygen. This of course raises the temperature of the combustion process, which is associated with increased concentrations of NO_x. Therefore, oxy-combustion is used simultaneously with the process of eliminating nitrogen from the air combustion (largely responsible for the formation of NO) and replacing it into exhaust gas (RFG - Recycled Flue Gas.). Despite the very high temperatures in the chamber (e.g. 1600 °C in the melting process), the concentration of nitrogen oxides may be lower, due to the elimination of nitrogen from combustion air. The primary obstacle to the propagation of oxy-combustion has so far been the high cost of obtaining pure oxygen. Since the pure oxygen production technologies have been improved and costs have been reduced, oxy-combustion can be applied in many industrial processes. Industry may be interested in this technology, where the conditions in the very high temperature contribute to the formation of large amounts of thermal NO. An additional advantage of the oxy-combustion is high combustion efficiency, lower volume of exhaust gases, less fuel consumption and therefore lower CO_2 and NO_x. Combustion of natural gas in O_2/CO_2 atmosphere allows optimising the combustion process. The results of experimental studies indicate that the combustion of natural gas in O_2/CO_2 with recycling exhaust gases has positive effects on reducing CO_2 emissions, a noticeable reduction or even elimination of NO_x and improve the efficiency of the furnace (Lampert & Ziębik, 2007; M.Wilk, Magdziarz & Kuźnia, 2010). It was noted that the major advantage of this technology is the ability to apply it in an existing energy plants (H.K. Kim, Kim, Lee & Ahn, 2007; Simpson & Simon, 2007).

Therefore, the problem seems to be both interesting and promising. The complex nature of the combustion process of natural gas causes the obtained experimental results which are not always repeatable so this issue requires further study.

2. Mechanisms of CO and NO_x formation in natural gas combustion processes

In many industries, because of the possibility of obtaining high temperatures, the primary fuel is natural gas, whose main component (ca. 98%) is methane. Combustion of natural gas is the source of the formation of many pollutants. During the combustion of natural gas in metallurgical furnaces the air pollutants are formed. These are nitrogen oxides, carbon oxides, and possibly trace amounts of hydrocarbons. The composition of natural gas varies slightly. The number and types of pollutants emitted from combustion are related to the composition of the fuel, the type of oxidation atmosphere used and the temperature prevailing in the combustion chamber.

2.1 The mechanism of CO formation

The relatively high CO emissions in combustion processes of natural gas occur in the following cases:

- Staging combustion for reduction of NO_x emissions,
- Inadequate mixing of air and fuel,
- Very rapid cooling of combustion products in the cold boundary layer of the combustion chamber.

The formation of CO in the flame is one of the main paths of reaction in the mechanism of combustion of hydrocarbons. Fuel hydrocarbons during the chemical degradation can be partially converted to CO. The formation of CO is done quickly, right at the beginning of the flame. The proposed overall reaction of formation of CO is as follows (for gaseous and liquid hydrocarbons) and volatiles of solid fuels. (Wilk, 2002; Bartok & Sarofim, 1991)

$$C_nH_m + \frac{n}{2}O_2 \rightarrow nCO + \frac{m}{2}H_2 \tag{1}$$

Oxidation of CO is strongly catalysed by even small amounts of hydrogen or its compounds with oxygen. There are two paths of CO oxidation. Main path of oxidation takes place at T > 1500 K (at p ≈ 0,1 MPa) and is as follows

$$CO + OH \rightarrow CO_2 + H \tag{2}$$

The second path CO oxidation takes place at T = 1000 - 1500 K and at p > 1 MPa is as follows

$$CO + HO_2 \rightarrow CO_2 + OH \tag{3}$$

HO_2 radical is produced in the recombination reaction:

$$H + O_2 + M \rightarrow HO_2 + M \tag{4}$$

HO_2 radical concentration is comparable to the concentration of OH radical reactions and rapid reactions (Bartok & Sarofim, 1991)

$$H + O_2 \leftrightarrow OH + O \tag{5}$$

$$O + H_2 \leftrightarrow OH + H \tag{6}$$

$$O + H_2O \leftrightarrow OH + OH \tag{7}$$

$$H + H_2O \leftrightarrow OH + H_2 \tag{8}$$

Competitive reactions to the above and CO oxidation reactions are the following recombination reaction

$$H + H \rightarrow H_2 \tag{9}$$

$$H + OH \rightarrow H_2O \tag{10}$$

$$H_2 + OH \rightarrow H_2O + H \tag{11}$$

H and OH radicals can meet together on the walls where it comes to their exhaustion, and it causes stopping of the CO oxidation reaction. In practice it is the result of too rapid cooling of exhaust gases below 1000 K. This is the main reason of no oxidation of CO to CO_2. The direct oxidation of CO by reaction:

$$CO + O_2 \rightarrow CO_2 + O \tag{12}$$

is very unlikely, because this reaction is very slow (the activation energy of this reaction is very high).

2.2 The mechanisms of NO formation

The primary adverse products of high temperature combustion are the nitrogen oxides NO_x. Knowledge of the mechanism of the NO_x formation can identify thermal and chemical conditions of furnaces and control of combustion processes, which affects prevention or reduction of the harmful substances emissions. The source of nitrogen oxides is nitrogen in the fuel and molecular nitrogen from the air. In combustion processes, there are two main types of nitrogen oxides: nitrogen monoxide, NO, and dioxide, NO_2. The main component of NO_x produced during natural gas combustion is NO, whose share in total NO_x emissions is typically at least 95%, and the rest is NO_2. The concentration of other oxides N_2O, N_2O_3 and N_2O_5 is low. The amount of NO_x in the exhaust gases depends mainly on the combustion temperature, excess air ratio and residence time in the reaction zone. There are four different mechanisms of the formation of NO:

- the thermal mechanism,
- the prompt mechanism,
- by means of N_2O
- the fuel NO_x.

During the combustion of natural gas, containing mostly methane, and not containing chemically bound nitrogen, the main way of NO_x formation mechanism in natural gas combustion is the thermal mechanism.

2.2.1 The thermal mechanism of NO formation

The thermal mechanism is based on the oxidation reactions of nitrogen from the air supplied for combustion, the rate becomes significant above 1400 °C. These reactions were first described by Zeldovich (Bartok & Sarofim, 1991; Warnatz et al., 2006; Tomeczek &, Gradoń, 1997; Muzio & Quartucy, 1997;. Flamme, 1998):

$$O + N_2 \xrightarrow{k_1} NO + N \quad k_1 = 1{,}8 \cdot 10^{14} \exp(-318 kJ \cdot mol^{-1} / (RT)) \, cm^3 / (mol \cdot s) \tag{13}$$

$$N + O_2 \xrightarrow{k_2} NO + O \quad k_2 = 6{,}4 \cdot 10^9 T \exp(-26 kJ \cdot mol^{-1} / (RT)) \, cm^3 / (mol \cdot s) \tag{14}$$

$$N + OH \xrightarrow{k_3} NO + H \quad k_3 = 3{,}8 \cdot 10^{13} \tag{15}$$

The term "the thermal NO" is connected with a very high activation energy due to a strong, triple-atomic bond in the molecule N_2. It is a highly endothermic reaction that runs with considerable speed only at temperatures higher than 1400 °C. The reaction of a hydrocarbon radical OH plays an important role in the combustion of humidified hydrocarbon fuels.

The rate of formation of NO is expressed by the equation (Warnatz et al., 2006; Gardiner, 2000)

$$\frac{d[NO]}{dt} = k_1[O][N_2] + k_2[N][O_2] + k_3[N][OH] \tag{16}$$

Atomic nitrogen is formed by the reaction (13) and is consumed in the reaction (14) and (15), hence the rate of formation is:

$$\frac{d[N]}{dt} = k_1[O][N_2] - k_2[N][O_2] - k_3[N][OH] \tag{17}$$

Taking into account the fact that reactions (14) and (15) are so fast that their products reach the equilibrium state, the preliminary assumption can be given:

$$\frac{d[N]}{dt} = 0 \tag{18}$$

and then equation (16) takes the form

$$\frac{d[NO]}{dt} = 2k_1[O][N_2] \tag{19}$$

k_1, k_2, and k_3 are the rate constants of the reaction.

The rate of formation of NO is controlled by the first, slow reaction of Zeldovich (13). If one molecule of one atom of NO and N is produced in this reaction, it immediately becomes the second particle produced by rapid reaction of NO (14). The formation of thermal NO takes place just behind the flame front in a zone of high temperatures (t > 1400 °C). The basic ways of reducing emissions of thermal NO in combustion processes are the reduction of the temperature, shortening the stay of the reagents in the zone of high temperatures and reducing the local concentration of N_2 and O_2.

Malte and Pratt proposed the mechanism of taking into account the role of N_2O in NO formation by the following reaction at temperatures lower than 1800 K (Kordylewski, 2008 (in Polish), Steele et al, 1995)

$$O + N_2 + M \rightarrow N_2O + M \tag{20}$$

$$N_2O + O \rightarrow NO + NO \tag{21}$$

$$N_2O + O \rightarrow N_2 + O_2 \tag{22}$$

These reactions, together with the reactions (13) and (14) in the literature are called "the extended thermal mechanism". The formation of NO by a mechanism of N_2O formation is particularly important at lower temperatures (T < 1200 °C) in the flames rather poor ($\lambda > 1$). Important role in the formation of N_2O plays the kind and characteristics of the third body M. It can be assumed that H_2O or its dissociation products (O, H, OH) affect the course of the reaction (20) (Wilk, 2002).

2.2.2 The prompt mechanism of NO formation

Fenimore conducting the experimental research on the combustion of rich mixtures (λ < 1) with various hydrocarbons (methane, ethane, propane) found that relatively high concentrations of NO occur in combustion zone just before the flame where the temperature does not exceed 750 °C. Fenimore was confident that there should be another mechanism for the generation of thermal NO and called him a "prompt" which is immediate. In the hydrocarbon flames there are not only O, H, OH radicals but also hydrocarbon radicals, where its highest concentration was found in the reaction zone of the flame. The hydrocarbon radicals are capable of activating N_2 reaction with the formation of nitrogen oxides in the flame. Fenimore assumed that CH_i hydrocarbon radicals react with nitrogen air according to the following reaction:

$$CH_2 + N_2 \rightarrow HCN + NH \tag{23}$$

$$CH + N_2 \rightarrow HCN + N \tag{24}$$

$$C + N_2 \rightarrow CN + N \tag{25}$$

Generally, these reactions can be written:

$$CH_x + N_2 \rightarrow HCN \text{ and other radicals } (CN, NH, N...) \tag{26}$$

The forming of amino and cyano compounds, among which the most important are HCN, NH, and CN, is oxidized to NO in the flame with the participation of radicals H, O, OH (Glarborg, Alzueta & Dam-Johansen, 1998). The prompt NO is formed very quickly during combustion. Velocity of its formation is like combustion velocity. The amount of formed NO depends weakly on temperature, but strongly depends on the local concentration of N_2. The NO prompt participates in further reactions running along the flame and lose their individuality.

2.2.3 The fuel NO mechanism

The amount of nitrogen in the fuel composition is very diverse. Nitrogen, in the gaseous fuel, is not chemically bonded with the combustible gas. However, it can occur as free molecular nitrogen N_2, which is the source of thermal or prompt nitrogen oxides. It is assumed therefore, that during the gas combustion the fuel nitrogen oxides do not occur.

2.2.4 The formation of NO$_2$

Miller and Bowman gave the most plausible explanation of the mechanism of formation of NO_2. They assumed that as a result of diffusion of H radicals from the flame in the area of low temperature (T < 750 °C) and high concentration of O_2 the reaction occurs (Miller & Bowman, 1989; Hori, 1986).

$$H + O_2 + M \rightarrow HO_2 + M \tag{27}$$

At the same time from the flame to low temperature zone NO diffuses, which comes in a rapid reaction with a peroxide radical reaction of HO_2 by:

$$NO + HO_2 \leftrightarrow NO_2 + OH \qquad (28)$$

In parallel, at higher temperatures, the NO_2 decomposition reactions may occur:

$$NO_2 + H \rightarrow NO + OH \qquad (29)$$

$$NO_2 + O \rightarrow NO + O_2 \qquad (30)$$

Therefore, under normal conditions of combustion (T = 1000 - 1700 °C, $\lambda \leq 1{,}3$) the final of NO_2 emission is low and does not exceed 5% of the total NO_x emissions. Significant impact on the formation of NO_2 next to a low temperature is connected with a high combustion pressure and the presence of hydrocarbons. The increase in the pressure favours the growth of the concentration of NO_2.

3. Experimental apparatus

The investigation of oxy-combustion process was conducted on a laboratory reactor containing a specially designed combustion burner, oxidizer preparation system, and temperature system, flow rate of combustion substrates control system, exhaust gas analysis system and exhaust gas system (Fig. 1).

The study included the characterization of the basic parameters of combustion, and above all took into account the effect of oxygen and carbon dioxide concentrations in the oxidizer on the exhaust gas composition and temperature profile along the combustion chamber.

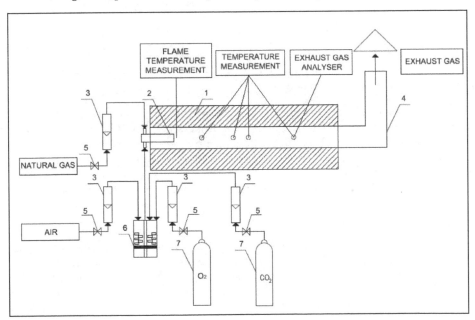

Fig. 1. Scheme of the experimental apparatus: 1 - combustion chamber, 2 - burner, 3 - rotameter, 4 - exhaust gas system, 5 - control valves, 6 - mixer, 7 - cylinder with oxygen, 8 - cylinder with carbon dioxide.

The specially designed kinetic burner, so called "pipe in pipe", was used. The inside diameter of burner was 34 mm and the outer diameter was 47 mm (Fig. 2). The combustion chamber was made from the heat-resistant steel with a length of 1310 mm and a diameter of 160 mm. Thermal isolation chamber was made of ceramic fibre with a thickness of 150 mm. Along the combustion chamber were holes, which allowed the measurement or analysis of the exhaust gas temperature inside the furnace. Oxidizer preparation system consisted of oxidizer pressure cylinders containing oxygen, air supply system with a fan and a mixer filled with the ceramic fittings, which enable more efficient mixing of the streams brought air and oxygen. The flow rates control system of combustion substrates included the rotameters and control valves. Mixing of fuel with an oxidizer took place in space between two pipes inside the burner, and was enforced by the system of holes in the outer pipe of the burner. The homogeneous mixture was introduced into the combustion chamber.

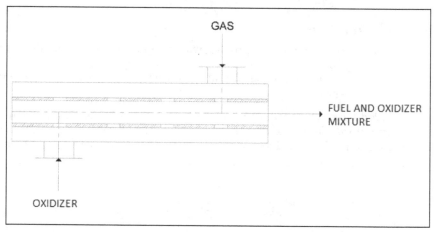

Fig. 2. Scheme of a specially designed burner.

To measure the flame temperature the thermocouple (PtRh-Pt) is installed in the wall of the chamber combined with a digital millimetre. The temperature profile along the length of the furnace was measured by NiCr-NiAl thermocouples at four points connected to a multichannel temperature recorder Czaki WRT-9 consisting of a microprocessor thermometer EMT 200, the switch places the PMP test. Concentrations of the combustion products (O_2, CO, CO_2, NO) were measured by the means of a gas analyser Testo 350 XL.

4. Results and discussion

The investigation of the oxy-combustion of natural gas is takes into account three types of oxidizing mixtures with an increased oxygen contents: 25% O_2, 27% O_2, 29% O_2. The parameters of the combustion process of the natural gas with the addition of oxygen to the combustion air are shown in Table 1. The study concerned the natural methane rich gas from the city with the following average composition: CH_4 - 98%, C_2-C_4 - 0,9%, N_2 - 1%, CO_2 - 0,1%.

To study the combustion of natural gas in modified atmosphere three options were carried out: the first - for the selected excess air ratio, the second - assumed a steady stream of gas

$\dot{V}_g$, m³/h	$\dot{V}_{oxidizer}$, m³/h	O_2, %	$\dot{V}_{CO_2}$, m³/h	T_{flame}, °C	$T_{exhaust\ gas}$, °C (in measured point)
0,8	6,3 - 9,5	21 - 29	0 - 2	1231 - 1316	885 - 1277

Table 1. The parameters of the combustion process of natural gas.

and air mixture for each oxygen-enriched air, and the obtained values λ resulted from the mixture of air and oxygen, and third option concerned the study of CO_2 addition to oxidizer whereas the air combustion was oxygen-enriched up to 27% and excess air ratio was: $\lambda_1 = 1,25$. The excess air ratio was calculated by taking into account the increased oxygen content in the mixture.

The first investigations were conducted for three values of the air excess ratio: $\lambda_1 = 1,15$, $\lambda_2 = 1,20$, $\lambda_2 = 1,25$. The flow of gas was constant $\dot{V}_g = 0,8$ m³/h. For each value of λ, by choosing an appropriate air and oxygen flows ratio, the air was oxygen-enriched in the range 21 - 29%. Table 2 and 3 shows the average values of measured experimental data.

The effect of oxygen addition to the combustion air on CO concentration for three different excess air ratios $\lambda_1 - \lambda_3$ was presented in Figure 3. For all the cases an increase in CO concentration with increasing oxygen concentration in the oxidising mixture was observed. The higher the excess air ratio, the lower was the concentration of carbon monoxide. It was observed that a small addition of oxygen around 4% slightly increases the concentration of CO, and the addition of 6% - 8% strongly increases the CO concentration. There was no

$\lambda_1 = 1,15$				
$\dot{V}_{air}$, m³/h	$\dot{V}_{O_2}$, m³/h	O_2, %	CO, ppm	NO, ppm
8,67	0	21	175	125
7,014	0,38	25	192	231
6,317	0,534	27	230	347
5,715	0,663	29	301	533
$\lambda_2 = 1,20$				
$\dot{V}_{air}$, m³/h	$\dot{V}_{O_2}$, m³/h	O_2, %	CO, ppm	NO, ppm
9,04	0	21	161	114
7,32	0,40	25	170	228
6,59	0,56	27	224	352
5,96	0,69	29	293	521
$\lambda_3 = 1,25$				
$\dot{V}_{air}$, m³/h	$\dot{V}_{O_2}$, m³/h	O_2, %	CO, ppm	NO, ppm
9,421	0	21	112	78
7,620	0,418	25	118	206
6,866	0,580	27	179	305
6,212	0,720	29	215	458

Table 2. The results of experimental studies of natural gas combustion in oxygen enriched atmosphere for $\dot{V}_g = 0,8$ m³/h, $\dot{V}_{CO_2} = 0$.

$\lambda_3 = 1{,}25; O_2 = 27\,\%$		
$\dot{V}_{CO_2}$, m³/h	CO, ppm	NO, ppm
0	179	305
0,5	182	144
1	210	73
1,5	242	43
2	355	24

Table 3. The results of experimental studies of natural gas combustion in CO_2 and oxygen enriched atmosphere for $\dot{V}_g$ = 0,8 m³/h, λ_3 = 1,25; O_2 = 27 %.

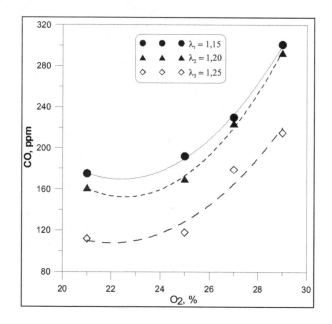

Fig. 3. The effect of oxygen addition to the combustion air on CO concentration in the natural gas combustion process with the oxygen enriched air depending on the excess air ratio.

difference in the concentration of CO depending on the excess air ratio used if air has been enriched to 29% O_2. It should be noted that the increase in O_2 concentration decreased the flow of oxidising mixture, so the observed increases of the concentrations of CO and NO are quite large. Figure 4 shows the effect of the oxygen addition to the combustion air for three different excess air ratios λ_1 - λ_3 of on the concentration of nitrogen oxide NO. Addition of oxygen to the oxidizer also causes an increase in NO concentration in the exhaust gas. NO concentration for all cases is at the same level.

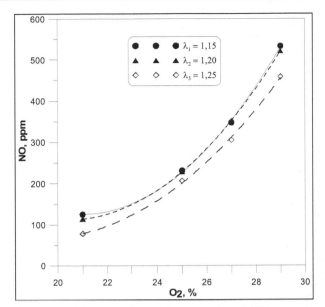

Fig. 4. The effect of oxygen addition to the combustion air on NO concentration in the natural gas combustion process with the oxygen enriched air depending on excess air ratio.

In the second series, the investigations were carried out for a fixed flow of gas and air in natural gas combustion in oxygen-enriched atmosphere. The combustion air was enriched from 21 to 29% oxygen, thereby generating the excess air ratios λ from 1,18 to 1,63. The results are presented in Table 3 and the graphs (Fig. 5 - 7).

λ	O_2, %	CO, ppm	NO, ppm	$T_{exhaust}$, °C
1,18	21	23	85	991
1,40	25	42	167	987
1,52	27	50	203	985
1,63	29	71	290	983

Table 4. The results of natural gas combustion in oxygen-enriched atmosphere for $\dot{V}_g$ = 0,8 m³/h and $\dot{V}_{air}$ = 9 m³/h, T_{flame} = 1140 - 1160 °C.

Figure 5 shows the effect of oxygen addition to the CO concentration in the combustion of natural gas in oxygen enriched air for constant flows of gas and air ($\dot{V}_g$ = 0,8 m³/h and $\dot{V}_{air}$ = 9 m³/h).

The increase of oxygen concentration in the oxidizer from 21 to 29% O_2, an increase in CO concentrations up to 300% was observed. Addition of oxygen in the oxidizer does not improve the complete combustion of the gas. CO molecule is more stable than CO_2 and its oxidation by oxygen is very slow, it is unknown whether pure carbon monoxide CO could be burnt (Kotowicz & Janusz, 2007). However, the oxidation of CO is possible by means of even a small concentration of hydrogen and its compounds, e.g. addition of a small concentration of water vapour would cause the CO combustion.

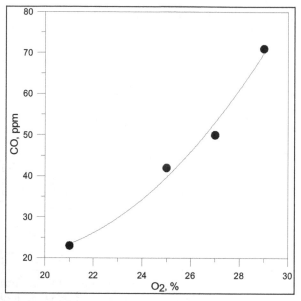

Fig. 5. The effect of oxygen addition on the CO concentration in the natural gas combustion in oxygen-enriched oxidizer for $\dot{V}_g$ = 0,8 m³/h and $\dot{V}_{air}$ = 9 m³/h.

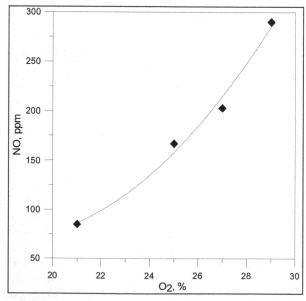

Fig. 6. The effect of oxygen addition on the NO concentration in the natural gas combustion in oxygen-enriched oxidizer for $\dot{V}_g$ = 0,8 m³/h and $\dot{V}_{air}$ = 9 m³/h.

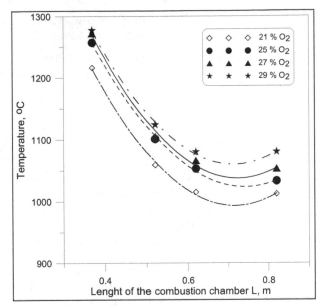

Fig. 7. The exhaust gas temperature profile along the length of the combustion chamber during combustion of natural gas in oxygen-enriched air.

The consequence of fuel combustion in oxygen-enriched atmospheres in O_2 + CO_2 system (replacing nitrogen by CO_2), is the reduction of pollutants emissions, mainly NO_x (Amann et. al, 2009; Li, Yan & Yan, 2009; Muskał et al., 2008). In the studied system (modified atmosphere: oxygen-enriched air combustion) lower concentrations of nitrogen oxide NO are not obtained, but on the contrary, more than threefold increase in NO concentration was observed (Fig. 6). Addition of oxygen increases the flame temperature, and therefore it also increases the NO concentration. Increased concentration of NO may also result from the larger concentration of oxygen in the reacting system, making easier the connection between the air nitrogen and oxygen at high temperature.

The temperature profile along the length of the furnace was also performed (Fig. 7). It was found that the exhaust gas temperature decreased along the furnace chamber in the measured points of the furnace, as well as an increased concentration of oxygen in the oxidizer. Oxygen enrichment of combustion air is done in order to raise the temperature of combustion in the furnace and to raise the growth rate of fuel combustion, which causes shortening of the flame. The reduction of the flame length explains the decrease of the exhaust gas temperature in the measured fixed points along the length of the furnace. The larger was the addition of oxygen, the flame was shorter and the temperature was lower in the test point. The flame temperature with the addition of oxygen increased from 1140 to 1160 °C.

In the course of the experiment it was observed that addition of oxygen resulted in visible changes in shape and colour of the flame, the bright crown of the burner nozzle and a very bright flame colour. Opportunity to observe these changes undoubtedly comes from reduction of the flame length. Adding oxygen to the combustion air also caused a change in the sound of the furnace operation to the louder, more intense one, associated with changes in the fuel and oxidizer mixture flow within the combustion chamber.

The study of natural gas combustion was also conducted in another modified atmosphere: $O_2/CO_2/N_2$. The process was studied under excess air ratio $\lambda_3 = 1{,}25$ and the oxidizer was oxygen-enriched up to 27%. CO_2 was added to the oxidizer in the range of 0 to 2 m^3/h.

The effect of carbon dioxide addition on the CO and NO concentrations of the studied process were investigated (Figure 8 and Figure 9). The CO concentration increased with CO_2 addition. The addition of 2 m^3/h of CO_2 has generated two times larger CO concentration comparing to the process operated in conventional atmosphere O_2/N_2. NO concentration, in contrary, decreased with increasing addition of CO_2. The maximum of CO_2 addition (2 m^3/h) decreased NO concentration ca. 12 times. The decrease of NO was connected with lower temperature obtained in the combustion chamber, because of large CO_2 thermal capacity. That fact confirms the flame temperature measured close to the burner nozzle in the axes of the flame presented in Figure 11. The decrease of flame temperature was observed with increasing CO_2 addition to the oxidizer. The thermal NO formation, main way of NO formation during natural gas combustion, takes place just behind the flame front in a zone of high temperature ($t > 1400\ °C$). Therefore, the efficient method of the NO reduction is the lower range of temperature used, shortening the stay of the reagents in the zone of high temperatures and reducing the local concentration of N_2 and O_2.

The exhaust gas temperature profile was conducted along the combustion chamber including the CO_2 addition (Figure 10). The exhaust gas temperature decreases with CO_2 addition along the combustion chamber. The more CO_2 is added the lower temperature is obtained.

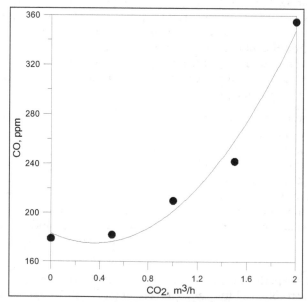

Fig. 8. The effect of carbon dioxide on the concentration of CO in the combustion of natural gas with oxygen-enriched oxidizer up to 27% O_2 and $\lambda_3 = 1{,}25$.

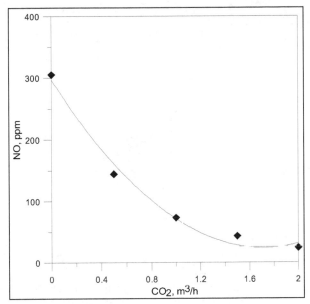

Fig. 9. The effect of carbon dioxide on the concentration of NO in the combustion of natural gas with oxygen-enriched oxidizer up to 27% O_2 and $\lambda_3 = 1{,}25$.

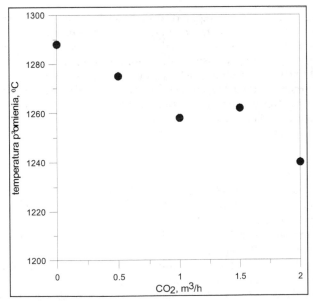

Fig. 10. The effect of carbon dioxide on the flame temperature in the combustion of natural gas with oxygen-enriched oxidizer up to 27% O_2 and $\lambda_3 = 1{,}25$.

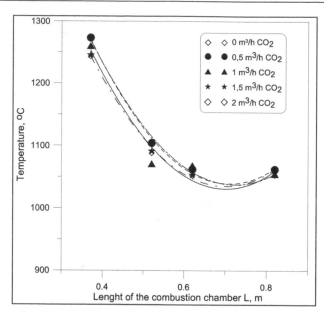

Fig. 11. The effect of carbon dioxide on the exhaust gas temperature measured along the combustion chamber in the combustion of natural gas with oxygen-enriched oxidizer up to 27% O_2 and $\lambda_3 = 1,25$.

5. Conclusion

In industrial processes, which are required to maintain very high temperatures such as in glass or in the production of cement, the oxy-combustion process can be used, or oxygen-enriched combustion air. This of course raises the temperature of the combustion process, which is associated with an increase in the concentration of NO_x. Therefore, the oxy-combustion should be used simultaneously with the process of elimination of nitrogen from combustion air (largely responsible for the formation of NO). Otherwise, the addition of oxygen increases the NO concentration, and therefore undesirable effect is achieved. A mixture of oxygen and carbon dioxide by replacing the air can lead to lower concentrations of nitrogen oxides by eliminating nitrogen from combustion air.

6. References

Amann J.-M., Kanniche M., Bouallou C. (2009). Natural gas combined cycle power plant modified into an O_2/CO_2 cycle for CO_2 capture, *Energy Conversion and Management*, Vol. 50, pp. 510-521,

Andersson K., Johnsson F. (2007). Flame and radiation characteristics of gas-fired O_2/CO_2 combustion, *Fuel*, Vol. 86, pp. 636-668,

Bartok W., Sarofim A. F. (1991). Fossil fuel combustion. A source book. *John Wiley and Sosns, INC*, New York, Chichester, Brisbane, Toronto, Singapore,

Buhre B.J.P., Elliott L.K., Sheng C.D., Gupta R.P., Wall T.F. (2005). Oxy-fuel combustion technology for coal-fired power generation, *Progress in Energy and Combustion Science*, Vol. 31, pp. 283-307,

Chen J.C., Liu Z.S., Huang J.S.: Emission characteristics of coal combustion in different O_2/N_2, O_2/CO_2 and O_2/RFG atmosphere, Journal of Hazardous Materials 2007, 142, s. 266-271,

Choi G., Katsuki M. (2001). Advanced low NO_x combustion using highly preheated air, Energy Conversion and Management, Vol. 42, pp. 639-652,

Croiset E., Thambimuthu K., Palmer A. (2000). Coal Combustion in O_2/CO_2 mixtures compared with air, The Canadian Journal of Chemical Engineering, Vol. 78, pp. 402-407,

Czakiert T., Bis Z., W. Muskała, Nowak W. (2006). Fuel conversion from oxy-fuel combustion in a circulating fluidized bed, Fuel Processing Technology, Vol. 87, s. 531-538,

Czakiert T., Nowak W., Bis Z (2008). Spalanie w atmosferach modyfikowanych tlenem, kierunki rozwoju dla kotłów CWF, Energia i Ekologia, Vol. 10, pp. 713-718 (in Polish),

Davidson J. (2007). Performance and cost of power plants with capture storage of CO_2, Energy, Vol. 32, pp. 1163-1176,

Flamme M. (1998): Low NO_x Combustion Technologies for High Temperature Application. Proceedings of 2nd International Symposium on Advanced Energy Conversion System and Related Technologies, (December 1-3), Nagoya, Japan, pp. 152-159,

Flamme M. (2001). Low NO_x combustion technologies for high temperature applications, Energy Conversion and Management, Vol. 42, pp. 1919-1935,

Gardiner W.C. (2000). Gas-Phase Combustion Chemistry. Springer-Verlag, New York,

Glarborg P., Alzueta M., Dam-Johansen K. (1998). Kinetic Modeling of Hydrocarbon/Nitric Oxide Interactions in a Flow Reactor. Combustion and Flame, Vol. 115, pp. 1-27,

Hori M. (1986). Experimental Study of Nitrogen Dioxide Formation in Combustion Systems. Twenty-first Symposium (International) on Combustion/The Combustion Institute/,pp. 1181-1188,

Kim H.K., Kim Y., Lee S.M., Ahn K. Y. (2007). NO reduction in 0.03-0.2 MW oxy-fuel combustor using flue gas recirculation technology, Proceedings of the Combustion Institute, Vol. 31, pp. 3377-3384,

Kordylewski W. (2008). Combustion and fuels, Oficyna wydawnicza Politechniki Wrocławskiej, Wrocław, Poland (in Polish),

Kotowicz J., Janusz K. (2007). Sposoby redukcji emisji CO_2 z procesów energetycznych, Rynek Energii, Vol. 1 (in Polish),

Lampert K., Ziębik A. (2007). Comparative analysis of energy requirements of CO_2 removal from metallurgical fuel gases, Energy, Vol. 32, pp. 521-527,

Li H., Yan J., Yan J., Anhenden M. (2009) Impurity impacts on the purification process in oxy-fuel combustion based CO_2 capture and storage system, Applied Energy, Vol. 86, pp. 202-213,

Miller J. A., Bowman C. T. (1989). Mechanism and Modeling of Nitrogen Chemistry in Combustion. Progress in Energy and Combustion Science, Vol. 15, pp. 287-338,

Muskał W., Krzywański J., Czakiert T., Sekret R., Nowak W. (2008). Spalanie w atmosferach modyfikowanych O_2 i CO_2, Energetyka, Vol. 10, pp. 669-671,

Muzio L.J., Quartucy G.C. (1997). Implementing NO_x Control: Research to Application. Progress in Energy and Combustion Science, Vol. 23, pp. 233-266,

Normann F., Andersson K., Leckner B., Johnsson F. (2008). High-temperature reduction of nitrogen oxides in oxy-fuel combustion, *Fuel*, Vol. 87, pp. 3579-3585,

Seepana S., Jayanti S. (2009). Flame structure and NO generation in oxy-fuel combustion at high pressures, *Energy Conversion and Management*, Vol. 50, pp. 1116-1123,

Simpson A., Simon A.J. (2007). Second law comparison of oxy-fuel combustion and post combustion carbon dioxide separation, *Energy Conversion and Management*, Vol. 48, pp. 3034-3045,

Steele R.C., Malte P.C., Nicol D.G., Kramlich J.C. (1995). NO_x and N_2O in Lean-Premixed Jet-Stirred Flames. *Combustion and Flame*, Vol. 100, pp. 440-449,

Szlęk A., Wilk R.K, Werle S., Schaffel N. Clean technologies generating energy from coal and the prospect of flameless combustion, *Rynek Energii*, Vol. 4, pp. 39-45 (in Polish),

Tan Y., Douglas M.A., Thambimuth K.V. (2002). CO_2 capture using oxygen enhanced combustion strategies for natural gas power plants, *Fuel*, Vol. 81, pp. 1007-1016,

Tomeczek J., Gradoń B. (1997). The Role of Nitrous Oxide in the Mechanism of Thermal Nitric Oxide Formation within Flame Temperature Range. Combustion Science and Technology, Vol. 125, pp. 159-180,

Warnatz J., Maas U., Diable R.W. (2006). Combustion. Physical and Chemical Fundamentals, Modeling and Simulation, Experimentals, Pollutant Formation. *4th Edition Springer-Verlag*, Berlin Heidelberg,

Wilk M., Magdziarz A., Kuźnia M. (2010). The influence of oxygen addition into air combustion on natural gas combustion process, *Rynek Energii*, Vol. 5, pp. 32-36 (in Polish),

Wilk R. (2002). Low-emission combustion. *Wydawnictwo Politechniki Śląskiej*, Gliwice, Poland,

Zhang N., Lior N. (2008). Two novel oxy-fuel power cycles integrated with natural gas reforming and CO_2 capture, *Energy*, Vol. 33, pp. 340-351.

8

Application of Natural Gas for Internal Combustion Engines

Rosli Abu Bakar[1], K. Kadirgama[1], M.M. Rahman[1], K.V. Sharma[1] and Semin[2]

[1]*Faculty of Mechanical Engineering, University Malaysia Pahang,*
[2]*Department of Marine Engineering,*
Institut Teknologi Sepuluh Nopember, Surabaya,
[1]*Malaysia*
[2]*Indonesia*

1. Introduction

It is well known that the fossil-fuel reserves in the world are diminishing at an alarming rate and a lack of crude oil is expected at the early decades of this century (Aslam et al., 2006). Gasoline and diesel fuel becomes scarce and most expensive (Catania et al., 2004). Alternative fuel becomes more conventional fuel in the coming decades for internal-combustion engines. Nowadays, the alternative fuel has been growing due to concerns that the reserves of fossil fuel all over the area are limited. Furthermore, the world energy crisis made the fossil-fuel price increases.

Natural Gas (NG) has been found in various locations in oil and gas bearing sands strata located at different depths below the earth surface (Catania et al., 2004). NG is a gaseous form of NG was compressed. It has been recognized as one of the promising alternative fuels due to its significant benefits compared to gasoline fuel and diesel fuel. These include reduced fuel cost, cleaner exhaust gas emissions and higher octane number. Therefore, the numbers of engine vehicles powered by NG were growing rapidly (Poulton, 1994; Pischinger, 2003). NG is safer than gasoline in many respects (Cho and He, 2007; Ganesan, 1999; Kowalewicz, 1984). The ignition temperature of NG is higher than gasoline fuel and diesel fuel. Additionally, NG lighter than air and dissipate upward rapidly. Gasoline fuel and diesel fuel will pool on the ground, increasing the risk of fire. NG is nontoxic and will not contaminate groundwater if failed. Advanced NG engines undertake significant advantages over the conventional gasoline engine and diesel engine (Kato et al., 1999). NG is a commonly available type of fossil energy. However, the investigation of applying NG as an alternative fuel in engines will be a beneficial activity, because the liquid fossil fuels will be finished and will become scarce and expensive (Catania, 2004; Sera, 2003). NG has some advantages compared to gasoline and diesel from the environmental perspective. It is a cleaner fuel than either gasoline or diesel as far as emissions are concerned. NG is considered to be an environmentally clean alternative to those fuels (Cho and He, 2007; Kato et al., 1999; Shashikantha and Parikh, 1999; Wayne, 1998). Advantages of NG as a fuel its octane numbers are extraordinarily suitable for spark ignition (SI) engines. NG engine can be operated in high compression ratio (Ganesan, 1999).

2. Natural Gas engine

2.1 Natural Gas engine development trend

There are four NG engine types, the traditional premixed charge spark ignition engine, the port injection lean burn engine, the dual-fuel/pilot injection engine, and the direct injection engine (Ouellette, 2000; Shashikantha and Parikh, 1999). Significant research has been done on these engines, the most promising of these, the injection engine requires further development in order to investigate the injection full potential. Shashikantha and Parikh (1999), studied a 17 kW, stationary, direct injection diesel engine converted to operate as a gas engine using producer-gas and NG as the fuels on two different operational modes called SIPGE (Spark Ignition Producer Gas Engine) and DNGE (Compressed NG Engine). Shashikantha and Parikh (1999) results of conversion to SIPGE (or DNGE) can be regarded as a success since comparable power and efficiency could be developed. NG operation of SIPGE yielded almost comparable power and higher efficiency, which establishes the fuel flexibility of the machine under spark ignition performance. The spark advance needed for producer-gas operation is much higher at 35° BTDC as compared to NG operation which is 22° BTDC, with compression ratio being same, i.e., 11.5:1.

Kato et al. (1999) has developed a new engine Toyota Camry that uses NG as fuel by modifying the base 2.2-liter gasoline engine in the unmodified engine, torque and power for NG decrease compared to gasoline. The new engine has adopted a high compression ratio, intake valves with early closed timing, intake and advanced exhaust valves with increased lift and a small back pressure muffler, which thereby restores the loss of engine power. Fig. 1 shows a multi port injection or multi point injection system was chosen by Czerwinski et al. (2003), and the injectors and pressure regulator have been recently developed in order to significantly reduce exhaust emissions. At the same time, precise air-fuel (A/F) ratio control and special catalysts for NG exhaust gas have been utilized. The resulting NG engines output power has been restored to approach that of the gasoline base engine. Wang and Watson (2000) have developed of a NG engine with ultra-lean-burn low emissions potential, hydrogen-assisted jet ignition (HAJI) is used to achieve reliable combustion and low NOx emissions, whilst direct injection is used to improve thermal efficiency and reduce hydrocarbon (HC) emissions. It is found that port-inducted propane, port-inducted NG and directly injected NG all produce negligible levels of CO and NOx.

The vast majority of NG engines in use today is premixed charge spark ignition engines (Chiu, 2004). Spark ignited (SI) engines have significant advantages over diesel engines in terms of particulate and NOx emissions, there are some drawbacks with respect to performance. Premixed SI engines allow 30% lower power output than equivalent size diesel engines due to knock limitations (Kato et al., 1999). In addition, SI engines receive high pumping losses, due to the need to throttle the intake air at part load conditions. These factors result in a 15 to 30% reduction in volumetric efficiency compared diesel engines (Brombacher, 1997). In diesel engine, Ouellette (2000) developed high pressure direct injection (HPDI) of NG in diesel engines, the result shown, that NG or methane are reduced by about 40% over diesel operation NOx. Peak torque loss 9% when running on NG compared to gasoline (Durell et al., 2000). Although peak power was not obtained on gas (due to the limitations of the injectors) there is also a predicted loss 9% on peak.

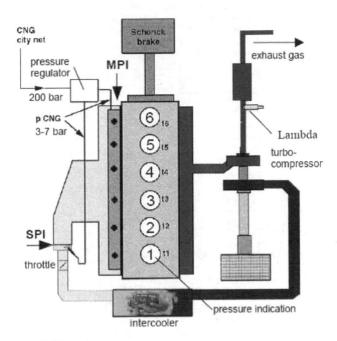

Fig. 1. Gas injection of NG engine.

The NG engine is best operated if such conditions as listed by Bakar et al. (2002) in Fig. 2. The principal operations are operated in high volumetric efficiency, turbulent flame speed, high compression ratio and proper air-fuel ratio. In operation in high volumetric efficiency and suitable of the air-fuel ratio is based on turbulent effect, injector type and lean burn operation. According to Bakar et al. (2002), injector is the important component in the best operation of NG engine. Another that, many researchers and institutions have contributed in improving the NG engine performance. In the area of increasing volumetric efficiency, Kubesh et al. (1995) developed an electronically controlled NG fueled engine with a turbocharged-aftercooled engine controlled by an electronic control system. Tilagone et al. (1996) found an increase up to 16% of thermal efficiency on a turbocharge spark ignition NG fuelled engine with multi point injection and optimized ignition timing with spark advance 200 higher running on stoichiometric A/F ratio.

In designing a turbulent effect in order to increase the flame speed combustion, Johansson and Olsson (1995) developed ten different geometries of the combustion chamber (CC). Their results showed a strong correlation between in cylinder turbulence and rate of heat release in the combustion process. However, the results also showed that geometries that gave the fastest combustion would gave the highest NOx values. In their further analysis, Johansson and Olsson (1995) developed six different CC to observe its effect on the combustion performance. The results showed different geometrical CC, with the same compression ratio (12:1), have extremely different combustion performance. The Quartette type of CC gave the highest peak turbulence. A squish-generated charge motion combustion chamber had its effect to the burning rates. High levels of turbulent generated from the

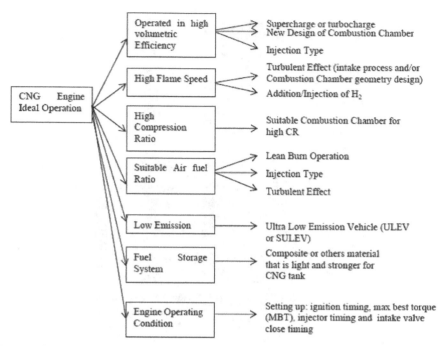

Fig. 2. The ideal NG engine operating conditions.

squish pretend to faster burning rates, which resulted in improvement of thermal efficiency. Evan et al. (1996) proved that the faster burning rates led to an average of 1.5% reduction in brake specific fuel consumption (BSFC) or 1.5% increase in power output under wide open throttle condition, as compare to the slowest burning cases. However the highest turbulence intensity combustion chamber also showed the highest emission.

In optimizing to the NG engine performance, Duan (1996) proposes the modification of setting up MBT, higher compression ratio and the use of gaseous fuel injection systems. Meanwhile, Ford introduces the NG Vehicle (NGV) truck by modifying fuel storage, fuel metering and emission control system. The injector timing, fuel control, spark advance, and exhaust gas recirculation (EGR) were also changed (Vermiglio, 1997). The simulations areas also conducted to increase the performance of NG engine. Oullette (1998) had simulated the combustion process and provides a better understanding of the injection and combustion process of the pilot-ignited directly-injected NG. The numerical simulation was expected to optimize the injection process by looking in especially at the geometry and the injection delay between two fuels. The model includes modifications for under expanded NG jets and includes a turbulent combustion model.

2.2 Injection methods of Natural Gas engine

There are four methods to inject the NG into the engine cylinder (Zastavniouk, 1997). First type is gas mixer / carburetor injection, second type is the single point injection, third type is multi point injection and fourth type is direct injection. The illustration of the four methods of NG injection is shown in Fig. 3.

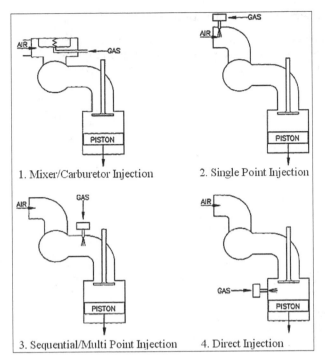

Fig. 3. Injection methods of NG engine.

The existing metering and mixing of the fuel may be accomplished using either a mechanical gaseous fuel mixer or carburetor, or an electronically controlled gaseous fuel metering system. This approach strives to achieve a homogeneous mixture of air and fuel before the air flow splits in the intake manifold. As discussed by Klimstra (1989), failure to obtain a homogenous mixture at this point can cause significant cylinder-to-cylinder variations in the air-fuel ratio. According to Zastavniouk (1997), Klimstra (1989) and Lino et al. (2008), this injection option can be increases emissions and the possibility of knock phenomena. Single point injection is use gaseous fuel injector to mix the gaseous fuel with the intake air in the manifold at one location for all cylinders of the engine. In this case, fuel is injected in a single location much like a gas mixer or carburetor. Single point electronic injection offers the advantage of more precise control of the amount of gaseous fuel entering the intake charge of the engine as well as the economy of using a minimum number of injectors (Zastavniouk, 1997). Multi point injection (MPI) is to inject the fuel into the each cylinder via intake port before intake valve (Czerwinski, 1999; 2003; Zastavniouk, 1997). This system uses one or more fuel injectors for each cylinder intake port of an engine and allows the designer to remove the fuel supply from the air supply area of the intake manifold. Direct injection is to inject the gaseous fuel directly into each combustion chamber of the engine.

In the MPI methods of NG, it is necessary to develop considerable turbulence during the compression stroke to obtain adequate air-fuel mixing. A high-turbulence, high swirl combustion chamber and high air-fuel mixing are useful for this type of injection to increase the engine performance.

2.3 Multi point injection system of Natural Gas

According to Lino et al. (2008), the main elements of the NG injection system are a fuel tank storing high pressure gas, a pressure reducer, a common rail and electro-injectors. The MPI system of NG is shown in Fig. 4.

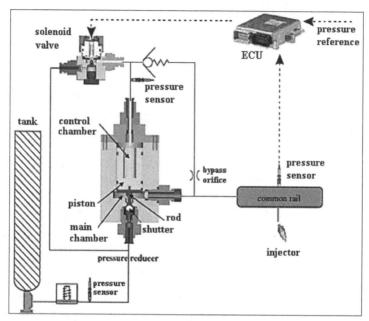

Fig. 4. Multi point injection system of NG engine.

The fuel coming from the tank supplies the pressure reducer before reaching the common rail and feeding the electronically controlled injectors. By supplying gas to the intake manifolds, injectors lead to the proper air/fuel mixture (Lino et al., 2008; Czerwinski et al., 1999; 2993). The large volume of the common rail helps in damping the oscillations due to the operation of both pressure controller and injectors. Namely, combining the electronic control of rail pressure with optimum design of the rail volume reduces the pressure oscillations inside the rail and leads to a more accurate fuel metering. The flow rate depends only on the rail pressure. Hence, the injected fuel quantity can be metered acting on rail pressure and injection timings are driven by the electronic control unit (ECU).

2.4 Diesel engine convert to multi point injection Natural Gas engine

In the diesel engines converted to run on NG, there are two main options discussed. The first is dual-fuel engine and the second is NG engine. Dual-fuel engine is referred to diesel engines operating on a mixture of NG and diesel fuel. NG has a low cetane rating and is not so suited to compression ignition, but if a pilot injection of diesel occurs within the gas/air mixture, standard ignition can be initiated. Between 50% and 75% of conventional diesel consumption can be replaced by gas when operating in this mode. The engine can also revert to 100% diesel operation. NG engines are optimized for the NG fuel. They can be derived from gasoline engines or may be designed for the purpose. According to Poulton (1994), until manufacturer original equipment (OE) engines are more readily available, however, the practice of converting diesel engines to spark ignition will continue, which involves the replacement of diesel fuelling equipment by a gas carburetor and the introduction of an ignition system and spark plugs. For compression ignition engines conversions to spark ignition, the pistons modified to reduce the actual compression ratio and a high-energy ignition system fitted (Czerwinski et al., 1999; 2003). The system is suitable for NG and is ideally suited to MPI system but can also be used for single point and low pressure in-cylinder injection. Gas production provides greater precision to the timing and quantity of fuel provided, and to be further developed and become increasingly used to provide better fuel emissions (Poulton, 1994).

The port injection NG produces negligible levels of CO, CO2 and NOx (Suga et al., 2000). In order to significantly reduce exhaust gas emissions, a port injection system was chosen by Czerwinski et al. (1999, 2003), Hollnagel et al. (1999, 2001) and Kawabata and Mori (2004), and the injectors and pressure regulator have been recently developed. In the same time, precise air-fuel (A/F) ratio control and specific catalysts NG exhaust gas has been utilized. By using it, NG engines output power near the gasoline base engine.

With the multi point injection, a high-speed gas stream is pulsed from the intake port through the open intake valve into the combustion chamber, where it causes effects of turbulence and charge stratification particularly at engine part load operations. The system is able to reduce the cyclic variations and to develop the border of lean operation of the engine. The flexibility of gas pulse timing offers the potential advantage of reduced emissions and fuel consumption. With three types of port injectors available on the market, Czerwinski et al. (2003) compared for stationary and transient engine operation. There are several advantages of port injection, e.g., better possibility to balance the air-fuel ratio of the cylinders, optimization of the gas injection timing and of the gas pressure for different operating conditions. The port injection has an injector for each cylinder, so the injectors can be placed in proximity to the cylinder's intake port. It also enables fuel to be delivered exactly as required for each individual cylinder (multi point injection) and enables more sophisticated technologies such as skip-firing to be used. Skip-firing is when only some of the cylinders are operating (the other cylinders are being skipped). This enables even more efficient use of the fuel at low loads, further lowering fuel consumption and unburned hydrocarbon (Czerwinski et al., 2003; Zastavniouk, 1997).

2.5 Multi point injection gas injector

In principle, the utilization of an optimal fuel-air mixture should provide the required power output with the lowest fuel consumption that is consistent with smooth and reliable operation (Zhao et al., 1995). Over past decades, MPI system has evolved into an electronic, pulse-width-modulated system that utilized multi point injectionly-timed separate injections into each intake port. According to Zhao et al. (1995), these transient sprays of 2.5 to 18.0 ms duration have a constant phase relative to the intake valve event, either for start on injection or end of injection, and provide significant advantages in engine transient response and hydrocarbon (HC) emissions. It is necessary to note, however, that the meteoric expansion in the use of such systems has generally out-paced the basic knowledge and understanding of the complex, transient fuel sprays that they produce.

According to Shiga et al. (2002), improvement of NG injector nozzle holes geometries and understand of the processes in the engine combustion is a challenge because the compression-ignition combustion process is unsteady, heterogeneous, turbulent and three dimensional and exceedingly complex. In MPI NG engines, NG is injected by fuel nozzle injector via intake port into the combustion chamber and mixing with air must occur before ignition of the gas fuel. To improve the perfect of the mixing process of NG fuel and air in the combustion chamber is arranging of nozzle holes geometry, nozzle spray pressure, modified of the piston head, arranging of piston top clearance, letting the air intake in the formation of turbulent and changing the NG fuel angle of spray (Mbarawa et al., 2001). The NG fuel spraying nozzle is the amount of earning variation so that can be done by research experimentation and computational of engine power, cylinder pressure, specific fuel consumption and missions which also the variation of them. Czerwinski et al. (1999, 2003) has researched the multi point injection injection of NG offers several advantages to increase the NG engine performance. The injector multi holes geometries development is to provide optimum fuel air mixing of the engine that will promote a similar engine performance (Ren and Sayar, 2001). According to Czerwinski et al. (2003) NG MPI has advantages for the more efficiency. The power, fuel consumption and thermal efficiency of the engine are higher than carburetor and single point. In the port injection NG engine, every cylinder has least one injector and the fuel are injected from the intake manifold into the engine cylinder when the intake valve is opened.

3. Development of multi point injection Natural Gas engine

The development of MPI NG engine is using diesel engine as a baseline engine. The fuel in the diesel engine is changed to NG. The ignition system is compression ignition changed to spark ignition. The fuel injection system is from direct injection mechanical system changed to multi point injection system and managed by electronic control unit. The NG engine is using throttle to control the intake air. The development of NG engine is reducing the compression ratio by modified the piston surface. The fuel is injected from the intake manifold into the engine cylinder when the intake valve is opened. The engine performance investigation is based on experimental and computational.

4. Multi point injection Natural Gas engine performance

4.1 Introduction

This chapter is exploring the engine performance based on experimental and computational. The engine computational model is used in the preliminary design to simulate the compression ratio effect of multi point port injection NG engine converted from diesel engine. The compression ratio has given the significant impact on engine power performance. In the engine computational model, if the compression ratio of the diesel engine convert to multi point port injection NG engine is designed in 12.5:1, 13.5:1, 14.5:1, 15.5:1, 16.5:1, 17.5:1, 18.5:1, 19.5:1 and 20.28:1, the brake power of the engine has been reduced 42.2%, 41.71%, 41.37%, 41.51%, 41.43%, 41.48%, 41.78%, 42.0% and 42.23%. Based on this brake power performance reduction effect from the compression ratio, the compression ratio with lower reduce brake power will be used in the engine conversion. The compression ratio 14.5:1 will be used in the development of multi point port injection NG engine. The engine conversion data are shown in Table 1.

Engine Parameter	Diesel Engine	NG Engine
Bore (mm)	86.0	86.0
Stroke (mm)	70.0	70.0
Displacement (cc)	407.0	407.0
Compression ratio	20.28:1	14.5:1
Ignition system	Compression Ignition	Spark Ignition
Engine Management	Mechanical Control	Electronic Control
Fuel system	Direct Injection	Multi point Port Injection
Fuel	Diesel	Natural Gas

Table 1. Specification the engine conversion.

4.2 Cylinder pressure of multi point injection Natural Gas engine

The results of cylinder pressure performance of direct injection diesel engine, compression ratio modified direct injection diesel engine and multi point port injection NG engine are shown in Fig. 5.

The results investigation of engine cylinder pressure is based on crank angle degree. The negative 180 to 0 crank angle degree is the engine compression stroke and the 0 to 180 crank angle degree is the engine power stroke for original diesel engine (ODE), compression ratio modified diesel engine (14.5CR DE) and MPI NG engine (NGE).

The engine cylinder pressure profile investigation results are shown in Fig. 5 are shown that the cylinder pressure is increasing in compression stroke to combustion ignition in crank angle negative180 degree bottom dead center (BDC) until around in crank angle 0 degree top dead center force (TDCF). In the compression stroke, the air-fuel volume is compressed from BDC to TDC. The simulation and experiment results are not similar. The simulation results are higher than the experimental results. The deviation is in average 2% for NG engine (NGE) and original diesel engine (ODE). The compression ratio of original direct injection diesel engine is 20.28:1, the compression ratio of modified direct injection diesel engine is 14.5:1 and the compression ratio of port injection NG engine is 14.5:1. From 1500 to 4000 rpm engine speed are shown that the original direct injection diesel engine cylinder pressure is higher than the modified direct injection diesel engine and multi point port injection NG engine. The highest of cylinder pressure is around in crank angle 0 degree (TDCF). From the cylinder pressure performance can be predicted that the product of engine power from the air-fuel combustion of original direct injection diesel engine is higher than modified direct injection diesel engine and the multi point port injection NG engine. The original direct injection diesel engine cylinder pressure is higher than modified direct injection diesel engine and multi point port injection NG engine because the compression ratio of original diesel engine is higher.

The highest of maximum cylinder pressure in the combustion process both of diesel engines and for multi point port injection NG engine are shown in Fig. 5a. In the original diesel engine, the maximum cylinder pressure is 84.0 bar declared in 1500 rpm engine speed. In the modified diesel engine, the maximum cylinder pressure is 61.1 bar declared in 1500 rpm engine speed. In the multi point port injection NG engine, the maximum cylinder pressure is 76.23 bar and declared in 1500 rpm engine speed. In this operating condition, both of diesel engines and NG engine combustion process are most excellent than the other condition. In the diesel engine, the 1500 rpm engine speed condition is not higher and not lower for the combustion of diesel fuel. Burned diesel fuel rate in 1500 rpm is most excellent to product the higher pressure and power. In the multi point port injection NG engine, the 1500 rpm engine speed condition is not higher and not lower for the combustion of NG engine. Burned NG fuel rate in 1500 rpm is most excellent and product the higher pressure and torque of the engine. The trend of the maximum cylinder pressure for original direct injection diesel engine, modified direct injection diesel engine and port injection NG engine are decrease if the engine speed is increased.

Fig. 5f shows the lowest of maximum cylinder pressure of direct injection diesel engine, modified direct injection diesel engine and MPI NG engine. The lowest maximum cylinder pressure in combustion process of original direct injection diesel engine, modified direct injection diesel engine and port injection NG engine are shown in 4000 rpm engine speed and the nominal is 72.82 bar for original diesel engine, 52.29 bar for modified diesel engine and 25.00 bar for port injection NG engine. In this case the combustion of diesel engines and NG engine are in lately so the combustion process is not excellent and unburned fuel is highest, this phenomenon can be decreasing the engine cylinder pressure performance. The port injection NG engine maximum cylinder pressure is lowest because the natural gas fuel is lower in density, hydrocarbon and energy than the diesel fuel. So, the cylinder pressure in the same compression ratio, the NG engine is lower than modified diesel engine if the engines are operated in high speed. The lowest cylinder pressure of original diesel engine is higher than modified diesel engine because the compression ratio of original diesel engine is higher than modified diesel engine.

The maximum cylinder pressure effect of the diesel engine converted to multi point port injection NG engine in the similar or higher compression ratio and in variation engine speed is shown in Fig. 5. In the 1500 rpm engine speed, the conversion of diesel engine to NG engine is increase the maximum cylinder pressure 8.97 %. In the 2000 rpm engine speed, the conversion of diesel engine to NG engine is decrease the maximum cylinder pressure 1.70 %.

In the 2500 rpm engine speed, the conversion of diesel engine to NG engine is decrease the maximum cylinder pressure 13.53 %. In the 3000 rpm engine speed, the conversion of diesel engine to NG engine is decrease the maximum cylinder pressure 39.12 %. In the 3500 rpm engine speed, the conversion of diesel engine to NG engine is decrease the maximum cylinder pressure 51.40 %. At the 4000 rpm, the conversion of diesel engine to NG engine is decrease maximum cylinder pressure 58.56 %.

The maximum cylinder pressure for NG engine is lower than the original diesel engine. It caused the compression ratio of NG engine is lower than the original diesel engine and the combustion energy output of diesel fuel is produces highest power than the natural gas fuel. Another that, the density of natural gas fuel is lower than the diesel fuel. So, in the same volume, the diesel fuel is has higher pressure than the gas fuel. In this engine conversion, the NG engine better to operate at low speed. In the low speed the maximum cylinder pressure increasing is higher dramatically than at the medium and high speed. For all of engine speed, the conversion of modified diesel engine to NG engine is increase the cylinder pressure in low speed, but in the high speed the engine conversion can be decreasing the cylinder pressure. In the high speed NG engine, the fuel energy is reduced and the combustion is not completely, but in the low speed the combustion of NG engine is completely because the combustion ignition is assisted by spark plug system and the

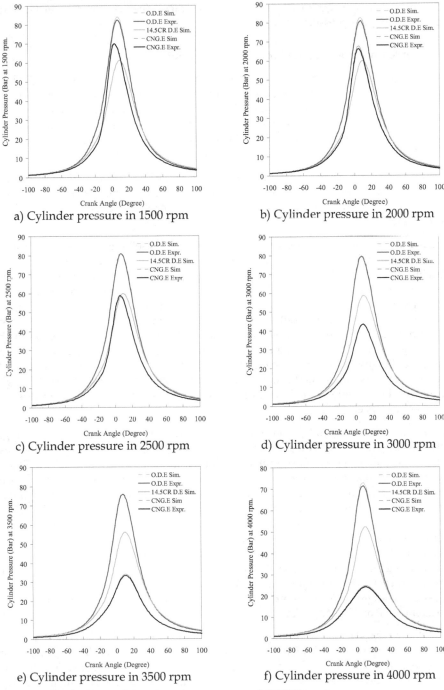

a) Cylinder pressure in 1500 rpm

b) Cylinder pressure in 2000 rpm

c) Cylinder pressure in 2500 rpm

d) Cylinder pressure in 3000 rpm

e) Cylinder pressure in 3500 rpm

f) Cylinder pressure in 4000 rpm

Fig. 5. Cylinder pressure of diesel engine convert to MPI NG engine.

ignition point of natural gas fuel is higher than the diesel fuel, so it can be producing the higher engine cylinder pressure.

4.3 Cylinder temperature of multi point injection Natural Gas engine

The investigation results of the engine cylinder temperature characteristics of original direct injection diesel engine (O.D.E), modified direct injection diesel engine (14.5CR D.E) and multi point port injection NG engine (NG.E) are shown in Fig. 6. In these figures, negative 180 to 0 degree is compression stroke and the 0 to 180 degree is power stroke for diesel engines and NG engine. The average deviation result of simulation and experiment is 2% for ODE and NGE.

In the low speed, the engine cylinder temperature of NG engine is higher than original diesel engine and modified diesel engine as shown in Fig. 6. In the high speed, the engine cylinder temperature fro both of diesel engines are higher than NG engine as shown from Fig. 6a to Fig. 6f.

The results are shown that increasing engine speed of diesel engine can be increase the maximum temperature in-cylinder engine. Unfortunately, the increasing engine speed of NG engine will be decrease maximum temperature in-cylinder engine. The decreasing engine speed of diesel engines will be decrease maximum temperature in-cylinder engine. Decreasing engine speed of NG engine will be increase maximum temperature in-cylinder engine. In this investigation results are shown that the highest maximum in-cylinder temperature in combustion process is not declared in the highest engine speed. In the both of diesel engines, the highest maximum temperature in-cylinder is declared in 3500 rpm engine speed, because in this case the combustion is most excellent than the other condition and unburned fuel is lowest, so the temperature product from the combustion is the highest. In the both of diesel engines, the lowest maximum temperature in combustion process is in 1500 rpm engine speed.

In this engine speed, the combustion process is not excellent and unburned fuel is highest than the other condition for compression stroke of compression ignition diesel engines. In the NG engine, the highest maximum temperature in-cylinder is declared in 1500 rpm engine speed, because in this case the combustion is most excellent than the other condition and unburned fuel is lowest, so the temperature product from the combustion is the highest. In the NG engine, the lowest maximum temperature in combustion process is in 4000 rpm engine speed. After 1500 rpm, the increasing engine speed, the combustion process is not excellent, the gas fuel density is lower, the air-fuel volume is lower and unburned fuel is highest than the other condition for compression stroke of natural gas spark assisted combustion engine. In the low speed, the engine cylinder temperature of NG engine is higher than original diesel engine and modified diesel engine as shown in Fig. 6a. In the high speed, the engine cylinder temperature fro both of diesel engines are higher than NG engine as shown from Fig. 6a to Fig. 6f.

The results are shown that increasing engine speed of diesel engine can be increase the maximum temperature in-cylinder engine. Unfortunately, the increasing engine speed of NG engine will be decrease maximum temperature in-cylinder engine. The decreasing engine speed of diesel engines will be decrease maximum temperature in-cylinder engine. Decreasing engine speed of NG engine will be increase maximum temperature in-cylinder

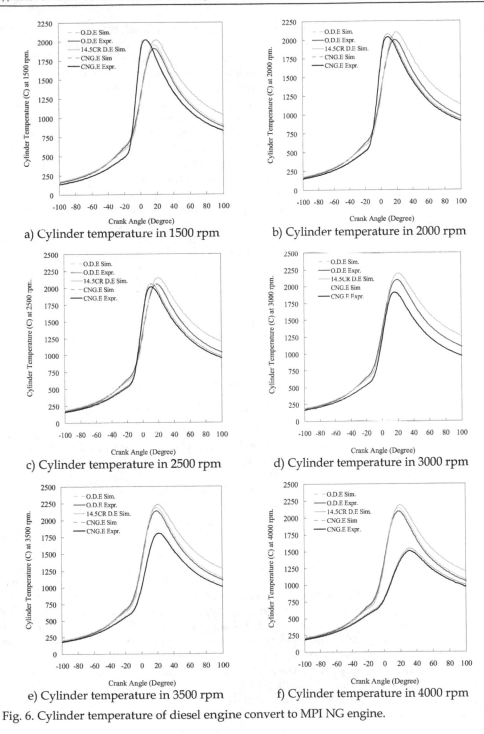

a) Cylinder temperature in 1500 rpm

b) Cylinder temperature in 2000 rpm

c) Cylinder temperature in 2500 rpm

d) Cylinder temperature in 3000 rpm

e) Cylinder temperature in 3500 rpm

f) Cylinder temperature in 4000 rpm

Fig. 6. Cylinder temperature of diesel engine convert to MPI NG engine.

engine. In this investigation results are shown that the highest maximum in-cylinder temperature in combustion process is not declared in the highest engine speed. In the both of diesel engines, the highest maximum temperature in-cylinder is declared in 3500 rpm engine speed, because in this case the combustion is most excellent than the other condition and unburned fuel is lowest, so the temperature product from the combustion is the highest. In the both of diesel engines, the lowest maximum temperature in combustion process is in 1500 rpm engine speed. In this engine speed, the combustion process is not excellent and unburned fuel is highest than the other condition for compression stroke of compression ignition diesel engines. In the NG engine, the highest maximum temperature in-cylinder is declared in 1500 rpm engine speed, because in this case the combustion is most excellent than the other condition and unburned fuel is lowest, so the temperature product from the combustion is the highest. In the NG engine, the lowest maximum temperature in combustion process is in 4000 rpm engine speed. After 1500 rpm, the increasing engine speed, the combustion process is not excellent, the gas fuel density is lower, the air-fuel volume is lower and unburned fuel is highest than the other condition for compression stroke of natural gas spark assisted combustion engine.

The effect of diesel engine converted to multi point port injection NG engine on the maximum engine cylinder temperature is shown in Fig. 6a – Fig. 6f. In the 1500 rpm, conversion of modified diesel engine to NG engine has been increase the maximum engine cylinder temperature 5.29% and 1.94 %. In the 2000 to 4000 rpm engine speed, conversion of engine has been decrease the maximum engine cylinder temperature 2.18%, 5.88%, 15.34%, 23.36% and 28.15%.

4.4 Torque performance of multi point injection Natural Gas engine

The engine torque performance investigation results of diesel engine convert to multi point port injection NG engine are shown in Fig. 7a – Fig. 7c.

The indicated torque results are shown in Fig. 7a. The simulation and experimental investigation results are not similar in 0.1 to 1.0 %. The simulation results are higher than the experimental results, it caused by the not excellently setting and reading data experiment and the assumption in simulation with not have losses. The indicated torque represents the thermodynamic work transferred from the gas to the piston converted via geometry to a torque applied to the crankshaft. In the original diesel engine, the highest indicated torque is 24.3453 Nm and declared at 3000 rpm engine speed. In the modified diesel engine, the highest indicated torque is 23.76 Nm and declared at 3000 rpm engine speed. The diesel engines indicated torque performance profile shows that from the minimum engine speed at 1500 rpm to 3000 rpm as the point of the highest indicated torque, the indicated torque performance is increase if the engine speed is increased until 3000 rpm engine speed. After 3000 rpm engine speed, the increasing engine speed can be decrease the indicated torque. In the port injection NG engine, the highest indicated torque is 20.4798 Nm and declared at 2000 rpm engine speed. After 2000 rpm engine speed, the increasing engine speed can be decrease the indicated torque. Based on Fig. 7a, the conversion diesel engine to NG engine can reduce the engine torque performance. The maximum indicated torque of diesel engine convert to NG engine is reduced 15.88%. The increasing engine speed can be increase the deviation percentage of the indicated torque of diesel engines compared to NG engine. On 1500 to 4000 rpm, the NG engine has reduce indicated torque of 11.08%, 13.43%, 23.51%,

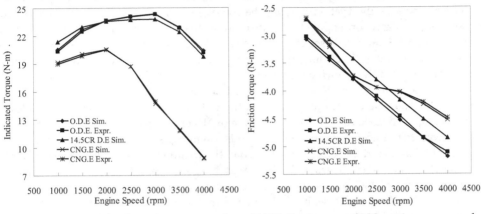

a) Indicated torque of NG engine compared to diesel engines

b) Friction torque of NG engine compared to diesel engines

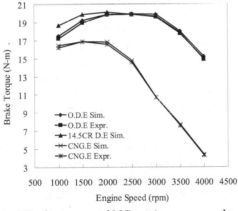

c) Brake torque of NG engine compared

Fig. 7. Torque performance of diesel engine convert to MPI NG engine.

41.29%, 50.34% and 56.54%. It meant that the thermodynamics energy were resulted from the diesel fuel combustion is higher than the NG fuel. It caused by the hydrocarbon chain, density and energy of diesel fuel of diesel engine is higher than the gas fuel of NG engine were ignited using spark assistant.

The friction torque result is sown in Fig. 7b. The simulation results are higher than the experimental results, it caused by the not excellently setting and reading data experiment and the assumption in simulation with not have losses. The highest friction torque in the original diesel engine is negative 5.18 Nm and modified diesel engine is negative 4.85 Nm declared at 4000 rpm engine speed. In the diesel engines, the friction torque is increase if the engine speed is increased. In the NG engine, the highest friction torque is negative 4.44 Nm and declared at 4000 rpm engine speed. In the NG engine and diesel engine, the friction torque is increase if the engine speed is increased. The conversion diesel engine to NG engine can increase the friction torque. The increasing engine speed can be increase the friction torque. Based on engine speed increasing, the percentage friction torque of NG engine is higher than the diesel engine in every engine speed. In 1500 rpm, the friction torque of NG engine is 15.02% and diesel engine is 15.21% from the indicated torque, where in this engine speed the reducing torque of diesel engine is higher than NG engine. In 2000 rpm, the friction torque of NG engine is 17.45% and diesel engine is 16.03% from the indicated torque, where in this engine speed the reducing torque of diesel engine is lower than NG engine. In 2500 rpm, the friction torque of NG engine is 20.57% and diesel engine is 17.29% from the indicated torque, where in this engine speed the reducing torque of NG engine is higher than diesel engine. In 3000 rpm, the friction torque of NG engine is 27.22% and diesel engine is 18.6% from the indicated torque, where in this engine speed the reducing torque of NG engine is exactly continue higher than diesel engine and the percentage deviation is increase. In 3500 rpm, the friction torque of NG engine is 36.4% and diesel engine is 21.23% from the indicated torque, where in this engine speed the reducing torque of NG engine is exactly continue higher than diesel engine and the percentage deviation is increase. In 4000 rpm, the friction torque of NG engine is 50.17% and diesel engine is 25.44% from the indicated torque, where in this engine speed the reducing torque of NG engine is exactly continue higher and the percentage deviation is increase. If the engine is running in higher than 1500 rpm engine speed, the NG engine friction torque is higher than diesel engine. In these cases, increasing engine speed will be increase the friction torque both of the engines, but the NG engine give the higher friction torque. It meant that the conversion of diesel engine to NG engine can be increase the friction torque of engine. It is caused by the natural gas as a fuel is less lubrication compared to diesel fuel as a liquid fuel and has the lubrication.

Brake torque of original diesel engine, modified diesel engine and NG engine are shown in Fig. 7c. The simulation results are higher than the experimental results, it caused by the not excellently setting and reading data experiment and the assumption in simulation with not have losses. Brake torque represents the torque available at the flywheel, after accounting for all friction and attachment losses as well as the acceleration of the crank train inertia. In the original diesel engine, the highest brake torque is 19.89 Nm declared at 2500 rpm engine speed. In the modified diesel engine, the highest brake torque is 20.12 Nm declared at 2000 rpm engine speed. In the diesel engine, the brake torque performance profile shows that from the low engine speed to medium engine speed, the brake torque performance is increase if the engine speed is increased. After 2500 rpm engine speed, the increasing engine speed can be decrease the brake torque. In the NG engine, the highest brake torque is 17.14 Nm and declared at 1500 rpm engine speed. After 1500 rpm engine speed, the increasing engine speed can be decrease the brake torque. The conversion of diesel engine to NG engine can reduce the engine brake torque. The maximum brake torque of modified diesel engine convert to NG engine is reduced 16.18%. The increasing engine speed can be increase the deviation point or percentage of the brake torque of diesel engine compared to NG engine. If the engines are run on 1500 to 4000 rpm, the conversion diesel engine to NG engine reduced brake torque 15.14%, 16.47%, 25.67%, 45.68%, 57.04% and 70.68%. Lower brake torque NG engine is caused by lower energy of and higher friction NG fuel.

4.5 Power performance of multi point injection Natural Gas engine

Indicated power performance of original diesel engine, modified diesel engine and port injection NG engine are shown in Fig. 8a. Indicated power represents the thermodynamic power transferred from the gas to engine. In the NG engine, the highest indicated power is 4.9 kW declared on 2500 rpm engine speed. From the minimum engine speed to 2500 rpm engine speed, the increasing engine speed can be increase the indicated power performance. After 2500 rpm to maximum engine speed, the increasing engine speed can be decrease the indicated power performance. The simulation and experimental investigation results are not similar in 0.05 to 0.5 %, it caused by the not excellently setting and reading data experiment and the assumed in simulation with not have losses. In the original diesel engine and modified diesel engine, the highest indicated power is 8.54 kW and 8.27 kW declared on 4000 rpm engine speed. The maximum indicated power of modified direct injection diesel engine convert to port injection NG engine is reduced 40.7%. The increasing engine speed can be increase the deviation point or percentage deviation of the indicated power of NG engine compared to diesel engine. From the 1500 to 4000 rpm, the conversion of modified diesel engine to NG engine can reduce indicated power 12.58%, 12.77%, 21.1%, 38.14%, 46.84% and 54.92%. It meant that the NG engine not applicable to run on high speed and very good power performance in medium speed. The combustion in high speed is produce lower power than the medium speed because the NG engine is developed in low compression ratio and thermodynamic energy. The effect of natural gas combustion of NG engine is lower in thermodynamic energy. The effect of lower thermodynamic energy is lower in indicated power.

Friction power of original diesel engine, modified diesel engine and port injection NG engine are shown in Fig. 8b. The simulation and experimental investigation results are not similar in 0.5 to 1.5 %. The simulation results are higher than the experimental results, it caused by the not excellently setting and reading data experiment and the assumption in simulation with not have losses. In the original and modified diesel engine, the highest friction power is negative 2.17 kW and 2.03 kW declared at 4000 rpm engine speed. The friction power profile shows that from the minimum engine speed at 1500 rpm to maximum engine speed on 4000 rpm, the friction power is increase if the engine speed is increased. In the multi point injection NG engine, the highest friction torque is negative 1.9 kW and declared on 4000 rpm engine speed. The friction power profile shows that increasing engine speed from the minimum engine speed at 1500 rpm to maximum engine speed at 4000 rpm is increase the friction power. In both of the NG engine and diesel engine, the friction power is increase if the engine speed is increased.

The conversion diesel engine to NG engine can increase the friction power performance percentage. The increasing engine speed can be increase the friction power of NG engine compared to diesel engine. The percentage friction power of NG engine is higher than the diesel engine in every engine speed. In 1500 rpm, the friction power of NG engine is 15.02% and diesel engine is 15.21% from the indicated power. In 2000 rpm, the friction power of NG engine is 17.45% and diesel engine is 16.03% from the indicated power. In 2500 rpm, the friction power of NG engine is 20.57% and diesel engine is 17.29% from the indicated power. In 3000 rpm, the friction power of NG engine is 27.22% and diesel engine is 18.6% from the indicated power. In 3500 rpm, the friction power of NG engine is 36.4% and diesel engine is

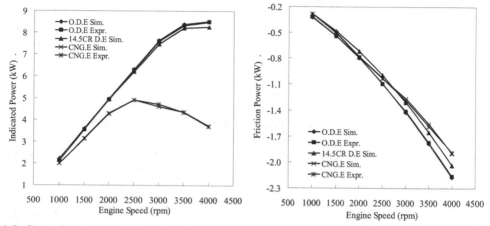

a) Indicated power of NG engine compared to b) Friction power of NG engine compared diesel engines   to diesel engines

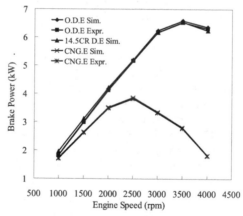

c) Brake power of NG engine compared

Fig. 8. Power performance of diesel engine convert to MPI NG engine.

21.23% from the indicated power. In 4000 rpm, the friction torque of NG engine is 50.17% and diesel engine is 25.44% from the indicated power. The percentage deviation of friction power compared to the indicated power of NG engine is lower than diesel engine in low engine speed until 1500 rpm. If the engine is running in higher than 1500 rpm engine speed, the NG engine friction power is higher than diesel engine. In these cases the increasing engine speed will be increase the friction power both of the engines and the NG engine give more friction power. It meant that the conversion of diesel engine to NG engine can be increase the friction power of engine. It is caused by the natural gas properties aspects, where in the natural gas as a fuel, the engine is less of the lubrication liquid compared to diesel fuel. Diesel fuel as a liquid fuel is have lubrication to reduce the friction.

Brake power performance of original diesel engine, modified diesel engine and port injection NG engine are shown in Fig. 8c. The brake power represents the power available at the flywheel, after accounting for all friction and attachment losses as well as the acceleration of the crank train inertia for brake torque. In the port injection NG engine, the highest brake power is 3.87 kW declared at 2500 rpm engine speed. In the NG engine, from 1500 to 2500 rpm, the increasing engine speed is increase brake power. After 2500 rpm engine speed, the increasing engine speed can be decrease the brake power. The simulation and experimental investigation results are not similar average in 1.0 %. The simulation results are higher than the experimental results, it caused by the not excellently setting and reading data experiment and the assumption in simulation with not have losses. In the original and modified diesel engine, the highest brake power is 6.6 kW and 6.54 kW declared at 3500 rpm engine speed. In the diesel engines, the brake power performance profile shows that from the minimum engine speed at 1500 rpm to 3500 rpm as the point of the highest brake power, brake power performance is increase if the engine speed is increased until 3500 rpm engine speed. After 3500 rpm engine speed, the increasing engine speed can be decrease the brake power. The conversion of four stroke direct injection diesel engine to multi point injection NG engine can reduce the engine brake power performance. The maximum brake power of direct injection diesel engine convert to multi point injection NG engine is reduced 41.374%. The increasing engine speed can be increase the deviation point or deviation percentage of the brake power of diesel engine compared to NG engine. So, the increasing engine speed can be increase the percentage deviation of brake power of the NG engine and diesel engines. If both of the engines are run on 1500 to 4000 rpm, the conversion of modified direct injection diesel engine to port injection NG engine has been reduced brake power 15.14%, 16.47%, 25.67%, 45.68%, 57.04% and 70.68%. The increasing engine speed can be increase the percentage deviation of power performance of NG engine compared to diesel engine. The reduction of the brake power in the NG engine is caused by low of brake torque. The low brake torque is caused by the low density of natural gas, low energy of natural gas, low volumetric efficiency, low flame speed, low compression ratio and higher friction of NG as an alternative fuel.

The investigation results on indicated power, friction power and brake power of the diesel engine converted to NG engine are shown that the engine conversion development can be decrease engine power performance. Increasing of engine speed over medium speed can be decreasing engine power performance of NG engine.

4.6 Fuel consumption of multi point injection Natural Gas engine

The simulation and experimental investigation results of the fuel consumption investigation of the diesel engines and NG engine are focuses on the indicated specific fuel consumption and brake specific fuel consumption. The investigation results are shown in Fig. 9.

Indicated specific fuel consumption (ISFC) of original diesel engine, modified diesel engine and port injection NG engine are shown in Fig. 9a. The ISFC is the nominal total fuel were used the engine to product their indicated power output. The simulation results are higher

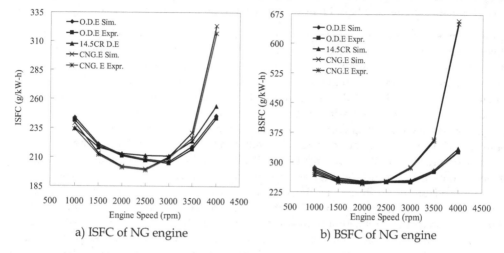

a) ISFC of NG engine
b) BSFC of NG engine

Fig. 9. Specific fuel consumption of diesel engine convert to MPI NG engine.

than the experimental results, it caused by the not excellently setting and reading data experiment and the assumption in simulation with not have losses. In the port injection NG engine, the fuel is entering to engine cylinder via intake port, the lowest ISFC is 199.593 g/kW-h declared on 2500 rpm engine speed and the highest ISFC is 323.532 g/kW-h

declared on 4000 rpm engine speed. From the minimum engine speed to 2500 rpm engine speed, the increasing engine speed can be decrease the ISFC. After 2500 to 4000 rpm, the increasing engine speed can be increase the ISFC. In the original and modified direct injection diesel engine, the fuel is injected directly to engine cylinder, the lowest ISFC are 205.93 g/kW-h and 210.98 g/kW-h declared on 3000 rpm engine speed. Then, the highest ISFC are 245.981 g/kW-h and 254 g/kW-h declared on 4000 rpm engine speed. The ISFC profile shows that from the minimum engine speed at 1500 rpm to medium speed, the increasing engine speed is decrease the ISFC. From medium speed to maximum engine speed at 4000 rpm, the increasing engine speed is increase the ISFC. The conversion of diesel engine to NG engine can be reduce the ISFC in the low to medium engine speed, but in the medium engine speed to high speed can be increase the ISFC. The minimum ISFC of multi point injection NG engine is reduced 2.9% compared to direct injection diesel engine. The maximum ISFC of port injection NG engine has increase 31.53% compared to direct injection diesel engine. The increasing engine speed can be increase the ISFC of NG engine compared to both of diesel engines. It meant that the NG engine not applicable to run on high speed, but it is very good ISFC in medium speed. The combustion in high speed is not excellent, not completely and high in unburned fuel, so the engine is produce lower power than the medium speed because the NG engine is developed in low energy and low density. The effect of the lower energy, density and power is increase the ISFC to the higher.

Brake specific fuel consumption (BSFC) of original diesel engine, modified diesel engine and port injection NG engine are shown in Fig. 9b. The BSFC is the nominal total fuel were used in the engine to product their brake power output. The simulation results are higher than the experimental results, it caused by the not excellently setting and reading data experiment and the assumption in simulation with not have losses. In the original and modified diesel engine, the lowest BSFC is 252 g/kW-h and 249.18 g/kW-h declared in 2500 and 2000 rpm engine speed. Then, the highest BSFC both of the diesel engines are 329.678 g/kW-h and 336.62 g/kW-h declared in 4000 rpm engine speed. In the original diesel engine, the BSFC profile shows that from 1500 rpm to 2500 rpm, the increasing engine speed can be decrease the BSFC. After 2500 rpm engine speed, the increasing engine speed can be increase the BSFC. In the modified diesel engine, the BSFC profile shows that from 1500 rpm to 2000 rpm, the increasing engine speed can be decrease the BSFC. But, after 2000 rpm engine speed, the increasing engine speed can be increase the BSFC. In the port injection NG engine, the lowest BSFC is 247.23 g/kW-h declared in 2000 rpm engine speed and the highest BSFC is 659 g/kW-h declared in 4000 rpm engine speed. From 1500 to 2000 rpm engine speed, the increasing engine speed of NG engine has been reduced the BSFC. After 2000 to 4000 rpm engine speed, the increasing engine speed can be increase the BSFC extremely. Its means, that the conversion of direct injection diesel engine to multi point injection NG engine can reduce the BSFC in lowest to medium engine speed and increase the BSFC in medium to highest engine speed. The minimum BSFC of modified direct injection diesel engine convert to port injection NG engine is reduced 0.78%. The maximum BSFC of modified direct injection diesel engine convert to port injection NG engine is increase 95.79%. The increasing engine speed can be increase the deviation point or deviation percentage of the BSFC of diesel engine compared to NG engine. In the low to medium engine speed, the unburned fuel in NG engine combustion is decrease so the out put can be product the higher torque and power. From the medium to the highest engine speed, the combustion of NG engine is not excellently, so the fuel consumption is increase because the unburned fuel is increase in the medium to highest speed. The effect of the increasing unburned fuel is can be decrease the engine brake torque and brake power. The effect of lower brake torque and brake power is increase the BSFC.

4.7 Mean effective pressure of multi point injection Natural Gas engine

The IMEP result is shown in Fig. 10a. The simulation and experimental investigation results are not similar average in 1.5 % for both of the engines. The simulation results are higher than the experimental results, it caused by the not excellently setting and reading data experiment and the assumption in simulation with not have losses. In the NG engine, the highest IMEP is 6.34952 bar declared on 2000 rpm engine speed. From the minimum engine speed to 2000 rpm engine speed, the increasing engine speed can be increase the IMEP. After 2000 rpm to maximum engine speed, the increasing engine speed can be decrease the IMEP. In the original and modified diesel engine, the highest IMEP is 7.5242 bar and 7.34 bar declared on 3000 rpm engine speed. The IMEP performance profile of diesel engines shows that from the minimum engine speed at 1500 to 3000 rpm engine speed, the increasing engine speed can be increase the IMEP, but from 3000 rpm to maximum engine speed at 4000 rpm, the increasing engine speed can be decrease the IMEP. Based on Fig.10a, the conversion diesel engine to NG engine can reduce the IMEP. The maximum IMEP of modified direct injection diesel engine convert to port injection NG engine has been reduced

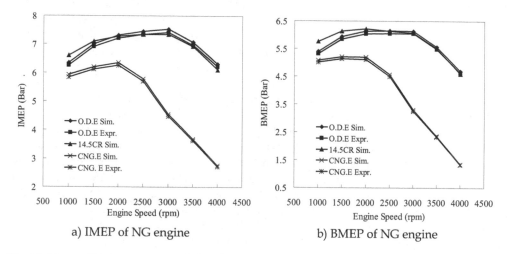

a) IMEP of NG engine   b) BMEP of NG engine

Fig. 10. Mean effective pressure of diesel engine convert to MPI NG engine.

13.54%. The increasing engine speed can be increase the deviation point or percentage deviation of the IMEP of NG engine compared to diesel engine. If the engine is run on 1500 to 4000 rpm, the conversion of modified diesel engine to port injection NG engine It meant that the NG engine not applicable to run on high speed and very good IMEP performance in low to medium speed. The combustion in high speed produces lower IMEP than the medium speed because the NG engine is developed in low energy. Effect of lower energy is causing the lower indicated torque. The lower indicated torque is can be reducing the lower IMEP.

Brake mean effective pressure (BMEP) performance of original diesel engine, modified diesel engine and port injection NG engine are shown in Fig. 10b. The BMEP is the external shaft work done per unit displacement. The simulation results are higher than the experimental results, it caused by the not excellently setting and reading data experiment and the assumption in simulation with not have losses. In the port injection NG engine, the highest BMEP is 5.21 bar declared on 1500 rpm engine speed. From 1500 to 1500 rpm, the increasing engine speed is increase BMEP. After 1500 rpm engine speed, the increasing engine speed can be decrease the BMEP. In the original and modified diesel engine, the highest BMEP are 6.148 bar and 6.22 bar declared at 2500 and 2000 rpm engine speed. The BMEP performance profile of original and modified diesel engine shows that the BMEP performance is increase if the engine speed is increased until 2500 and 2000 rpm engine speed. After 2500 and 2000 rpm engine speed, the increasing engine speed can be decrease the BMEP. The conversion of four stroke direct injection diesel engine to multi point injection NG engine can be reducing the engine BMEP performance. The maximum BMEP of modified direct injection diesel engine convert to port injection NG engine is reduced 16.18%. If the engines are run on 1500 to 4000 rpm, the conversion of diesel engine to port injection NG engine reduced BMEP 15.14%, 16.48%, 25.67%, 45.68%, 57% and 70.68%.

The investigation results on mean effective pressure performance such as indicated and break mean effective pressure of direct injection diesel engine converted to port injection NG engine are shown that the engine conversion development can be decrease the engine power performance. The reduction of the mean effective pressure in the NG engine is caused by lower energy, density and higher friction of compressed natural gas as an alternative fuel for engines. The increasing of engine speed over the medium speed can be decrease the engine mean effective pressure performance of NG engine. The highest of engine mean effective pressure of NG engine is declared in medium engine speed and the lowest mean effective pressure is declared in highest engine speed. The increasing engine speed will be increase the percentage deviation of mean effective pressure performance of NG engine compared to diesel engine.

5. Conclusion

The original diesel engine cylinder pressure is higher than the modified diesel engine and CNG engine. It caused the compression ratio of NG engine is lower than the original diesel engine and the combustion energy output of diesel fuel has to produce the highest power than natural-gas fuel. Another that, the density of natural-gas fuel is lower than the diesel fuel. The increasing engine speed of the diesel engine has increase the maximum temperature in engine. Unfortunately, the increasing engine speed of NG engine has been decrease maximum temperature inin- engine. The decreasing engine speed of diesel engines has been decrease maximum temperature inin- engine. Decreasing engine speed of NG engine has increase maximum temperature in engine. Engine torque, power, mean effective pressure and efficiency performance of original direct injection diesel engine cylinder pressure is higher than the modified diesel engine and sequential port injection dedicated NG engine. It meant that the thermodynamics energies were resulted from the diesel fuel combustion is higher than the NG fuel. It caused by the hydrocarbon chain, density and energy of diesel fuel of the diesel engine is higher than the gas fuel of NG engine were ignited using to spark assistant. The fuel consumption of NG engine is higher than the diesel engine. Fuel consumption is increased because the unburned fuel is increased in the medium to the highest speed of NG engine. The effect of the increasing unburned fuel is can decrease the engine brake torque and brake power. The effect of lower brake torque and brake power increase the brake specific fuel consumption..

6. References

Aslam, M.U.; Masjuki, H.H.; Kalam, M.A.; Abdesselam, H.; Mahlia, T.M.I.; Amalina, M.A. (2006). An experimental investigation of CNG as an alternative fuel for a retrofitted gasoline vehicle. *Fuel* 85: 717–724.

Bakar, R. A.; Sera, M.A.; Mun, W.H. (2002). Towards The Implementation of CNG Engine: A Literature Review Approach To Problems And Solutions. *Proc. of BSME-ASME International Conference on Thermal Engineering.* December 31, 2001 – January 2, 2002. Dhaka. BSME-ASME. 2002.

Brombacher, E.J. (1997). Flow Visualisation of Natural Gas Fuel Injection, *MSc Thesis*, University of Toronto.

Catania, A.E.; Misul, D.; Spessa, E.; Vassallo, A. (2004). Analysis of combustion parameters and their relation to operating variables and exhaust emissions in an upgraded multivalve bi-fuel CNG SI engine. *SAE Paper*. 2004-01-0983.

Chiu, J.P. (2004). Low Emissions Class 8 Heavy-Duty, On-Highway Natural Gas and Gasoline Engine. *SAE Paper*. 2004-01-2982.

Cho, H. M.; He, Bang-Quan (2007). Spark ignition natural gas engines - A review. *Energy Conversion and Management* 48: 608–618.

Czerwinski J.; Comte P.; Janach.W.; Zuber P. (1999). Sequential Multipoint Trans-Valve-Injection for Natural Gas Engines. *SAE Paper*. 1999-01-0565.

Czerwinski, J.; Comte, P.; Zimmerli, Y. (2003). Investigations of the Gas Injection System on a HD-CNG-Engine. *SAE Paper*. 2003-01-0625.

Duan, S.Y. (1996). Using Natural Gas in Engines: Laboratory experience with the use of natural gas fuel in IC engines. *IMechE Seminar Publication*, 1996: 39–46.

Durrel, Elizabeth.; Allan, Jeff.; Law, Donald.; Heath, John. (2000). Installation and Development of a Direct Injection System for a Bi-Fuel Gasoline and Compressed Natural Gas. *Proceeding of ANGV Conference 2000*. October 17 – 19, 2000. Yokohama: ANGV.

Evan, R.L.; Blaszczyk, J.; and Matys, P. (1996). An Experimental and Numerical Study of Combustion Chamber Design for Lean-Burn Natural Gas Engines. *SAE Paper*. 961672.

Ganesan, V. (1999). *Internal Combustion Engines*. N.Delhi: Tata McGraw-Hill.

Hollnagel, Carlos.; Borges, Luiz.H.; Muraro, Wilson. (1999). Combustion Development of the Mercedes-Benz MY1999 CNG-Engine M366LAG. *SAE Technical Paper* 1999-01-3519.

Hollnagel, C.; Neto, J. A. M.; Di Nardi, M. E.; Wunderlich, C.; Muraro, W. (2001). Application of the natural gas engines Mercedes-Benz in moving stage for the carnival 2001 in Salvador City. *SAE Paper*. 2001-01-3824.

Johansson, B. and Olsson, K. (1995). Combustion Chambers for Natural Gas SI Engines Part I: Fluid Flow and Combustion. *SAE Paper*. 950469.

Kato, K.; Igarashi, K.; Masuda, M.; Otsubo, K.; Yasuda, A.; Takeda, K.; Sato, T. (1999). Development of engine for natural gas vehicle. *SAE Paper*. 1999-01-0574. 1999.

Kawabata, Y and Mori, D. (2004). Combustion diagnostics and improvement of a prechamber lean-burn natural gas engine. *SAE Paper*. 2004-01-0979.

Klimstra, J. (1989). Carburetors for Gaseous Fuels - On Air-to-Fuel Ratio, Homogeneity and Flow Restriction. *SAE Paper* 892141.

Kowalewicz, Andrzej. (1984). *Combustion System of High-Speed Piston I.C. Engines*. Warszawa: Wydawnictwa Komunikacji i Lacznosci..

Kubesh, J.T.; Podnar, D.J.; Gugliemo, K.H. and McCaw, D. (1995). Development of an Electronically-Controlled Natural Gas-Fueled John Deere Power Tech 8.1L Engine. *SAE Paper* .951940.

Kubesh, John. T.; Podnar, Daniel. J. (1998). Ultra Low Emissions and High Efficiency from On-Highway Natural Gas Engine. *SAE Paper* 981394.

Lino, Paolo.; Maione, Bruno.; Amorese, Claudio. (2008). Modelling and predictive control of a new injection system for compressed natural gas engines. *Control Engineering Practice*, 2008. 16 (10): 1216-1230.

Mbarawa, M.; Milton, B.E.; Casey, R.T. (2001). Experiments and modelling of natural gas combustion ignited by a pilot diesel fuel spray. *Int. J. Therm. Sci.* 40: 927–936.

Oullette, P.; Mtui, P.L.; Hill, P.G. (1998). Numerical Simulation of Directly Injected Natural Gas and Pilot Diesel Fuel in a Two Stroke Compression Ignition Engine. State of Alternative Fuel Technologies. *SAE Paper* 981400.

Ouellette, Patric. (2000). High Pressure Direct Injection (HPDI) of Natural Gas in Diesel Engines. Proceeding ANGVA 2000 Conference. Yokohama.

Pischinger, S.; Umierski, M.; Hüchtebrock, B (2003). New CNG concepts for passenger cars: High torque engines with superior fuel consumption. *SAE Paper*. 2003-01-2264.

Poulton, M.L. (1994). *Alternative Fuels for Road Vehicles*, Computational Mechanics Publication, London.

Ren, W and Sayar, H. (2001). Influence of nozzle geometry on spray atomization and shape for port fuel injector. *SAE Paper*. 2001-01-0608.

Sera, M.A.; Bakar, R.A.; Leong, S.K. (2003) CNG engine performance improvement strategy through advanced intake system. *SAE Paper*. 2003-01-1937.

Shashikantha.; Parikh, P.P. (1999). Spark ignition producer gas engine and dedicated compressed natural gas engine-Technology development and experimental performance optimization. *SAE Paper*.1999-01-3515.

Shiga, S.; Ozone, S.; Machacon, H. T. C.; Karasawa, T. (2002). A Study of the Combustion and Emission Characteristics of Compressed-Natural-Gas Direct-Injection Stratified Combustion Using a Rapid-Compression-Machine, *Combustion and Flame* 129:1–10.

Suga, T.; Muraishi, T.; Brachmann, T.; Yatabe, F. (2000). Potential of a natural gas vehicle as EEV. *SAE Paper*. 2000-01-1863.

Tilagone, R.; Monnier, G.; Chaouche, A.; Baguelin, Y. and Chauveron D.S. (1996). Development of a High Efficiency, low Emission SI-CNG Bus Engine. *SAE Paper*. 961080.

Vermiglio. E; Jenskins, T.; Kleliszewski, M.; Lapetz, J.; Povinger, B.; Willey, R.; Herber, J.; Sahutske, K.; Blue, M, and Clark, R. (1997). Ford's SULEV Dedicated Natural Gas Trucks. *SAE Paper*. 971662.

Wang, D. E and Watson, H. C. (2000). Direct injection compressed natural gas combustion and visualization. *SAE Paper* 2000-01-1838.

Wayne, W. S.; Clark, N. N.; Atkinson, C. M. (1998). A parametric study of knock control strategies for a bi-fuel engine. *SAE Paper*. 980895.

Zastavniouk, Oleg. (1997). Study of Mixing Phenomena in a Dual Fuel Diesel Engine Air Intake Manifold. *MSc Thesis*. University of Alberta.

Zhao, F.; Lai, M.; Harrington, D.L. (1995). The Spray Characteristics of Automotive Port Fuel Injection – A Critical Review. *SAE Paper*. 950506.

9

Defining a Gas Turbine Performance Reference Database Model Based on Acceptance Test Results

Norberto Pérez Rodríguez, Erik Rosado Tamariz,
Alfonso Campos Amezcua and Rafael García Illescas
Instituto de Investigaciones Eléctricas (IIE)
Mexico

1. Introduction

With the growing participation of natural gas (NG) in power generation industry, a global worry about the effects of chemical composition changes on the behavior of gas turbines. A comparison of fossil fuels consumption used in power industry is shown in Figure 1 [1]. This figure represents a projection for the years 2003-2013 according to what is expected by the Mexican electric industry.

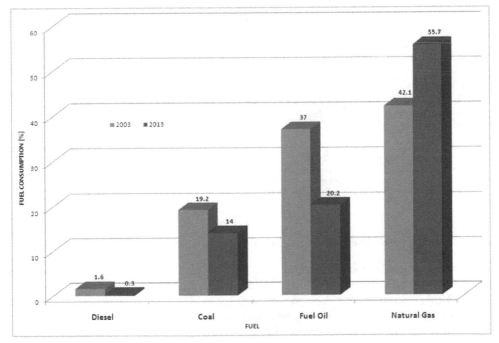

Fig. 1. Fuels Consumption in Mexican Power Industry.

The reduction of the fuel oil consumption observed in the previous figure is due to the process of substitution of this fuel by natural gas. This is because environmental reasons and techniques for incrementing the installed capacity through turbo gas plants and combined cycles based on natural gas.

As a sample of interest of the industry about this topic, it is possible to see that several technical papers related to this problematic, have been published during the last decade. All these works deal with gas quality, the variations on emissions, the increase in the noise levels and the consequences on gas turbine operation [2-10].

As departure point to evaluate the performance of turbogas power plants is analyzes, the increasing demand of electrical power in the world, as well as the necessity count with more modern and efficient electrical power plants orients to the design, installation and commercialization of new power plants which will be developed in harmony with the environment and promote energy saving. Like a behavior model of this phenomenon in Mexico is shown in Figure 2 where is presented the growth perspective to year 2016 of the national power sector in terms of generated gross energy.

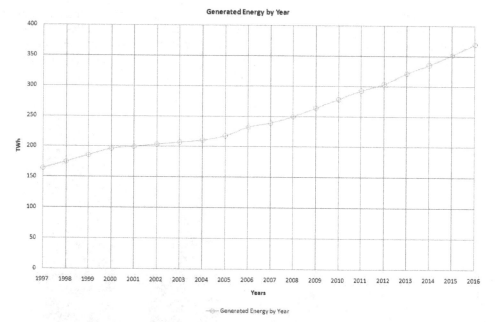

Fig. 2. Generated Gross Energy by Years in Mexico [11].

The growth expectation and modernization in terms of generation of electrical power in the central zone of Mexico gave rules to the Mexican Company of Electric Energy Generation (MCEEG) that operate in this zone to realize the acquisition of electrical power plants of distributed generation on strategic and critics zones around of Mexico city. Therefore MCEEG drove the installation of fourteen 32MW turbo gas electrical power plants impelled by aero derivatives turbines which burn natural gas.

As a result of the technical and administrative necessity of the MCEEG; combined with the effective national regulations in Mexico on terms of energetic production, it is realized collaboration on Mexican Electrical Research Institute (MERI) with the objective to develop this task during the works development until the commercial operation of each unit generation.

The main requirement to obtain the authorization certificate for beginning of the commercial operation of each one of fourteen power plants was the performance evaluation of each unit expressed in terms of average net power capacity, heat consumption rate, maximum noise levels and permissible levels of emissions. The MERI was to realize the behavior evaluation of these parameters through the determination of the average net power capacity, heat consumption rate, maximum noise levels and permissible levels of emissions from data collected during acceptance test of each turbo gas unit which was lead by a company specialized in the subject proposed by the supplier of the commissioning service of power plants and guaranteed by the MCEEG. As an additional product of the work carried out by the MERI, was develop a thermodynamics model based on real parameters obtained in each acceptance test which represents the initial operational state of the units and specifies the guidelines for a future trending monitoring system. The power plant has an aeroderivative turbine installed in simple cycle, operating with a cooling system (chiller) with effective capacity for a specific altitude and operation under any load regimen up to 32 MW.

2. Performance evaluation process

These tests are made on the final individual devices of each plant and integrate approaches to assessing the quality of the product. This is done in order to verify agreement between the guaranties and collected data on their performance under normal conditions of operation of the turbo unit, at different load levels with all its controls in automatic mode.

The goal of the tests is to verify if all devices of the open cycle turbo generator comply with the requirements established before, such as the guaranteed net capacity with the net unit heat consumption, guaranteed low level of noise and guaranteed pollutant emissions.

2.1 Power plant performance

The performance tests of Average Net Power Capacity (P_{NET}) and Heat Consumption Rate (Q_{RATE}) for 32MW turbo gas power plants were carried out in operation condition under base load with the air cooling system at turbine entrance running.

The measurements were developed with a combination of high precision instruments installed isolated and independent of the whole system (not connected to it). The measuring system implemented in the plant itself was also done. The measurements with high precision instruments were developed with instruments installed specifically for the performance test. Measurements of the plant are those made with permanent instrumentation of the plant, including measurements made by the flow meters from the gas station.

The net capacity (P_{NET}) was measured on the high side of the generator step-up transformer with the installed watt-hour meter (Digital-Multi-Meter DMM-B). The energy readings were recorded manually from the screen meter. The readings of energy (kWh) and the precise schedules (time) were used to calculate the average net capacity (P_{NET}) and net unit heat consumption (Q_{RATE}) carrying out the test using the following formulations:

$$P_{NET} = \frac{PM_{END} - PM_{START}}{\Delta t} \tag{1}$$

$$Q_{CONS} = \frac{m_r(Q_{ATM} + h_{CC} + h_{ATM})}{\Delta t} \tag{2}$$

$$Q_{RATE} = \frac{Q_{CONS}}{P_{NET}} \tag{3}$$

Where: PM_{END} and PM_{START} represented the power measurement at end and beginning the test (kWh), Δt is the time period of the test (h), Q_{CONS} is the heat consumption (kJ/h), m_r is the mass of fuel used during the test period (kg), Q_{ATM} is Reference specific energy of fuel (kJ/kg), h_{CC} and h_{ATM} represented the specific enthalpy of the fuel (combustion chamber) and the reference specific enthalpy of the fuel (kJ/kg).

These tests results of the P_{NET} and Q_{RATE} should be corrected considering the variation between the current environmental conditions at the time of the test and the normal environmental conditions. Those are listed in Table 1.

PARAMETER
Fuel
Equipment Status (Load)
Atmospheric Temperature
Relative Humidity
Chiller Temperature
Altitude
Barometric Pressure
Low Calorific Power (Fuel)

Table 1. Basis Conditions of the Tests.

In order to meet the guaranteed values of performance in each turbo gas unit, it was necessary to adjust the values of (P_{NET}) and (Q_{RATE}) considering the base conditions. This was done by means of the use of correction curves by temperature, barometric pressure, relative humidity and fuel calorific value.

The Corrected Average Net Power Capacity (CP_{NET}) and The Corrected Net Unit Heat Consumption (CQ_{RATE}) are calculated by correcting the P_{NET} and Q_{RATE} values to the guarantee basis conditions; this is done following the correction methodology shown below:

$$CP_{NET} = \frac{P_{NET}}{F_{1P} \cdot F_{2P} \cdot F_{3P} \cdot F_{4P}} \tag{4}$$

$$CQ_{RATE} = \frac{Q_{RATE}}{F_{1HR} \cdot F_{2HR} \cdot F_{3HR} \cdot F_{4HR}} \tag{5}$$

Where: F_{1P}, F_{2P}, F_{3P} and F_{4P} represented the correction factor of measured power versus atmospheric temperature, humidity, barometric pressure and low calorific power (fuel) for corrected net capacity. F_{1HR}, F_{2HR}, F_{3HR} and F_{4HR} represented the correction factor of measured power versus atmospheric temperature, humidity, barometric pressure and low calorific power (fuel) for corrected heat consumption rate.

The correction factors are calculated by linear interpolation between the data points shown in the correction curves (see Figures 3, 4, 5 and 6). The data measured in each run test are averaged before entering the calculation of interpolation.

2.2 Noise level emissions evaluation

In Mexico, noise contamination produced by utilities is regulated by the Secretary of the Environment and Natural Resources ("SEMARNAT" by its initials in Spanish) [12]. This office establishes the policies and applicable regulations. On the other hand, the Mexican Official Standard ("NOM" by its initials in Spanish) [13] is the responsible for establishing maximum limits allowable for noise emissions produced by each industrial or power plant inside or outside the facilities.

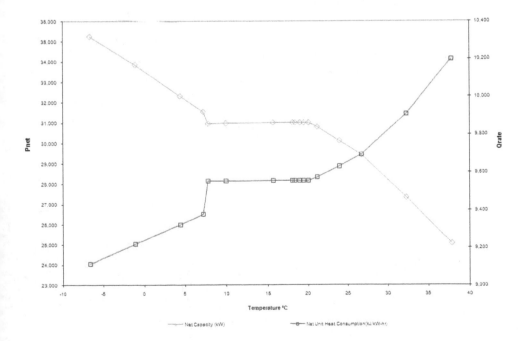

Fig. 3. Correction curve by temperature.

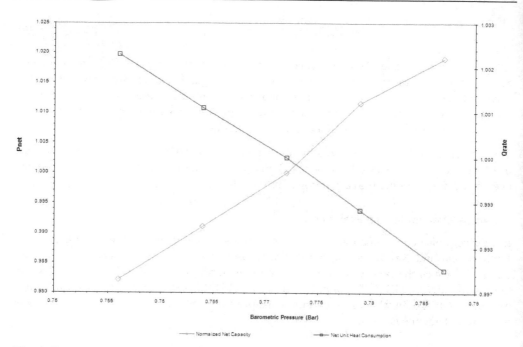

Fig. 4. Correction curve by barometric pressure.

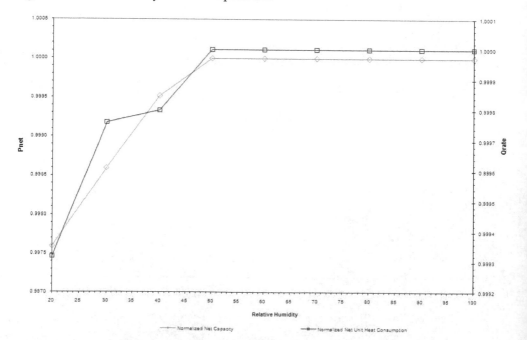

Fig. 5. Correction curve by humidity.

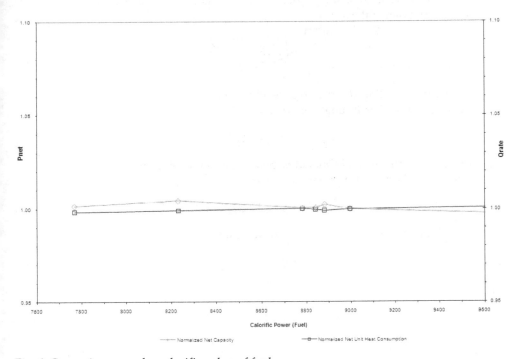

Fig. 6. Correction curve by calorific value of fuel.

In order to observe the previously described requirements, several activities were carried out during commissioning procedure of plants. Noise levels produced by some critical machines or equipments were obtained in near field (at the machine) and far field (around the facilities). All levels measured were below the comfort limits established by the standards, so they cannot reduce the quality of life of people living in the surroundings.

In order to get a correct measuring data processing acquired during far and near field noise evaluation, time intervals should be taken into account as well as minimum and maximum levels for each measuring point.

The equivalent sound level at each measuring point is determined through the mathematical expression (6):

$$N_{EQ} = 10\log\frac{1}{m}\left(\sum_m 10\frac{N}{10}\right) \tag{6}$$

Then the means must be calculated from the measured values for each point by means of equation (7):

$$N_{50} = \frac{\sum_i^n N_i}{n} \tag{7}$$

In a similar way, from equation (8) the numerical values of standard deviation can be assessed for each point of the test:

$$\sigma = \frac{\sqrt{\sum (N_i - N_{50})^2}}{n-1} \tag{8}$$

It is also important to find out the noise level values that were present during more than 10% of total recording time through the equation (9):

$$N_{10} = N_{50} + 1.28817\sigma \tag{9}$$

Then corrected numerical values should be obtained from previous measurements: Correction due to the presence of extreme values:

$$C_e = 0.9023\sigma \tag{10}$$

N_{50} average values from the fixed source and back noise:

$$\Delta_{50} = (N_{50})_{Source} (N_{50})_{Far} \tag{11}$$

Back noise correction:

$$C_f = -(\Delta_{50} + 9) + 3\sqrt{4\Delta_{50} - 3} \tag{12}$$

The numeric noise pressure value Nff from the noise source is obtained from equation (8):

$$N_{ff} = N_{50} + C_e \tag{13}$$

Finally, the last correction from back noise is used to determine the real numerical value of the noise source (equation 13)

$$N'_{ff} = N_{ff} + C_f \tag{14}$$

Where: N_{EQ} is the equivalent sound level (dBA), N is the fluctuant sound level (dBA), m is the total number of measurements, N_{50} is the mean of measured values (dBA), N_i is the sound level during certain time (dBA), n is the number of observations for each measuring point and σ is the standard deviation of measurements.

2.3 Emissions evaluation

Air quality and effects on the environment that produce the greenhouse gases emissions are of major importance for development of any country like Mexico. The nitrogen oxide emissions (NOx) from power generation equipment that uses technologies based on combustion cycles; are strictly regulated in order to develop monitoring and verification process of its generated atmospheric emissions during operation, according to Mexican Official Standards (NOM) which are compatible with the International Standards. For that previous reason, and based on the geographic location of the power plant facilities and according to Mexican Official Standards [14], this site is denominated like a critical zone (C_{ZN}) in Mexico because of high atmospheric pollution levels. This demonstrates the importance of establishing an emissions reference model that the plant generates during operation process.

The reference model was generated from the emission diagnosis develop to the power plant during the performance test at the end of the commissioning procedure. This model will be used as a reference basis for evaluation of emissions averages rates that the unit must handle during the history of operation.

Emissions testing and measurements were accomplished when the unit was operating at base load (100%) and at partial load of 75% with a 15% of O2.

The equipment used during testing consists of a fixed CEMS (Continuous Emission Monitoring System) which was calibrated prior and posterior to tests. This is done by comparing to a certified composition of reference gases found to be within the allowable deviation of +/- 3% [15].

Measurements were recorded manually from PC interface of the emission monitoring. Emissions were stored and endorsed every 110 minutes for both cases (full and partial load) observing a stable behavior of emissions.

After NOx measurements, some statistical calculations with the obtained data were done determining the mean and standard deviation for each load and observing the behavior of such values during load changes.

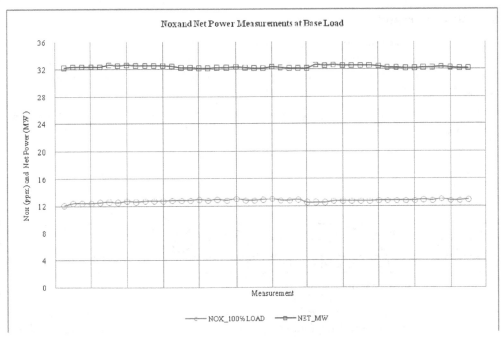

Fig. 7. NOx measurements (ppm) and Net Power (MW) recorded during test at 100% load.

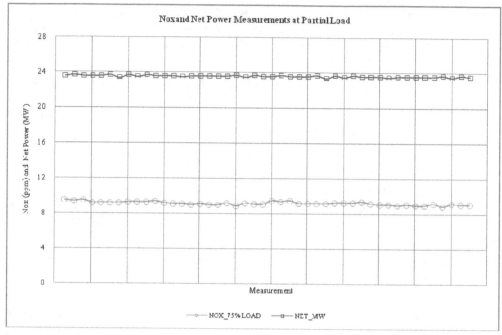

Fig. 8. NOx measurements (ppm) and Net Power (MW) recorded during test at 75% load.

3. Evaluation results

Once the tests, measurements, partial calculations and corrections of the mathematical models have been completed, it is come to make an evaluation of the results obtained with respect to the reference values guaranteed by the manufacturer and with the applicable standards for each parameter.

Firstly the results evaluation on terms of Average Net Power Capacity (P_{NET}) and Net Unit Heat Consumption (Q_{RATE}) at base load and partial load (75 percent) it is realized.

According to the values provided by the manufacturer and based on the average basic operations conditions for each one of the fourteen power plants, the guaranteed reference values of Average Net Power Capacity (P_{NET}) and Net Unit Heat Consumption (Q_{RATE}) are shown in the table 2:

PARAMETER	GUATANTEE VALUE	UNITS
Guaranteed Net Capacity	31,368	kW
Guaranteed Net Unit Heat Consumption	9,545	kJ/kWh

Table 2. Performance guarantees.

A comparison summary on term of Guaranteed Average Net Power Capacity with respect to the obtained results in test for fourteen power plants at base load and partial load are shown in Figures 9 and 10.

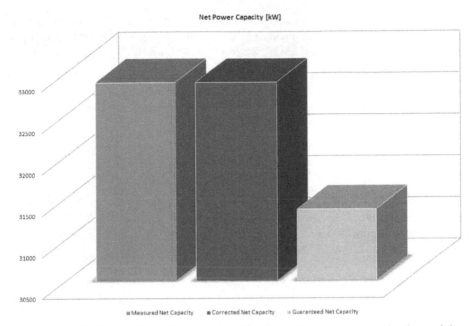

Fig. 9. Average Net Power Capacity comparison test results vs. guaranteed values of the units at base load.

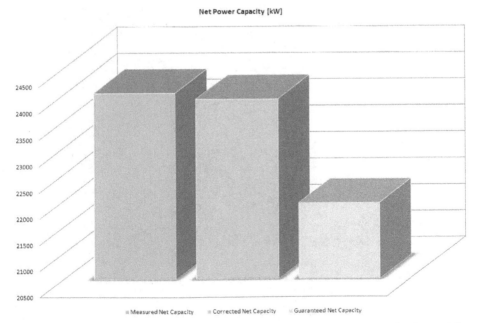

Fig. 10. Average Net Power Capacity comparison test results vs. guaranteed values of the units at partial load (75 percent).

Similar form to a the previous summary now appears a comparison in terms of Guaranteed Net Unit Heat Consumption with respect to the obtained results in test for fourteen power plants at base load and partial load are shown in Figures 11 and 12.

In the other hand the NOx measurements, some statistical calculations with the obtained data were done determining the mean and standard deviation for each load (100% and 75%) and observing the behavior of such values during load changes. Figure 13 shows graphically the recorded deviations.

Figure 14 shows a comparison of reference values established by the national standards [15] an international standards [16], the guaranteed value by the manufacturer [17] and the mean values obtained during the tests.

Once obtained the near and far field noise pressure levels during the test for the fourteen power plants, they were compared to the reference value found in the Mexican standard [13] as shown in Figure 15. From that standard it can be seen that the reference value is higher than the measured one guaranteed by the manufacturer.

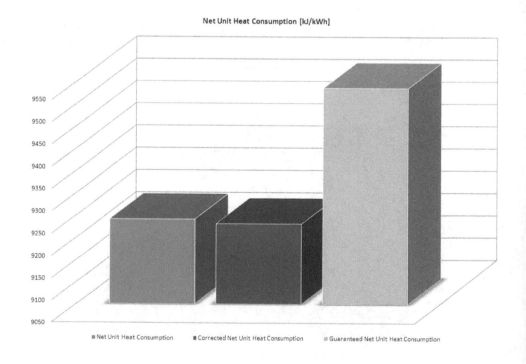

Fig. 11. Net Unit Heat Consumption comparison test results vs. guaranteed values of the units at base load.

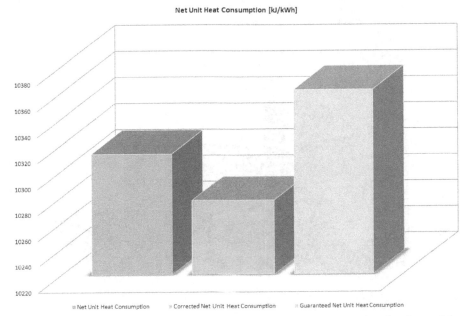

Fig. 12. Net Unit Heat Consumption comparison test results vs. guaranteed values of the units at partial load (75 percent).

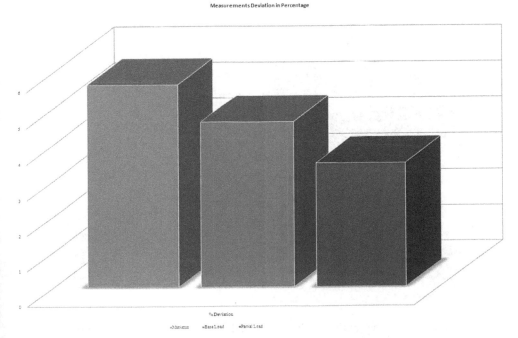

Fig. 13. NOx measurement deviation in percentage obtained during tests.

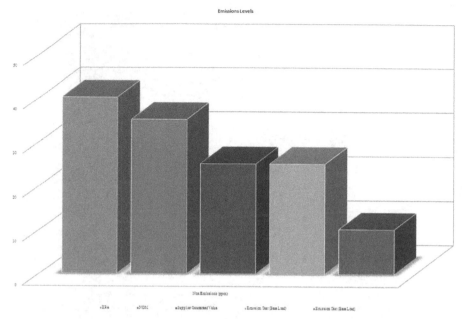

Fig. 14. Comparison of the results obtained during the tests with regard to national and international standards and guaranteed values by the manufacturer.

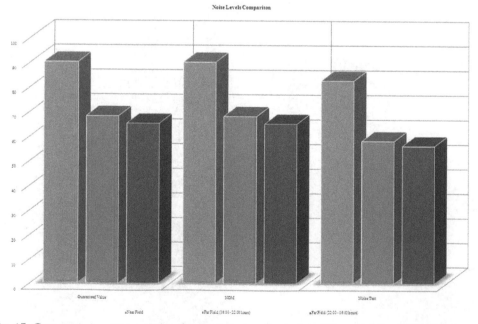

Fig. 15. Comparison of the results obtained during the tests with regard to national and guaranteed values by the manufacturer.

4. Reference model

The reference model based on acceptance test results were generated from the diagnosis develops to the power plants during performance test at the end of the commissioning procedure. This model will be used as a reference basis for the evaluation averages rates that the unit must handle during the history of operation.

In figure 16 is show behavior tendency in terms of average net capacity, developed from the calculated values during the performance test for each one of the fourteen power plants. That is behavior and variation rates that must be have any unit during history operation to maintain it is energetic efficiency.

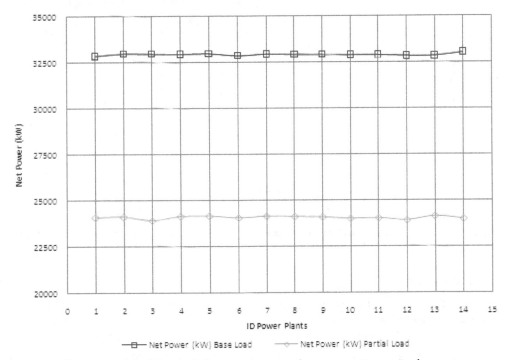

Fig. 16. Reference model of the variation rate in term of average net capacity for a power plant unit at base load or partial load (75 percent).

Furthermore is shown the performance of the unit in terms of emissions to determinate the grade of air pollution exhaust for the power plants. This reference model also was developing from the calculated values during the performance test for each one of the fourteen power plants. It is show in Figure 17.

As a final result of the reference model developed its obtains a Mathematical expressions that consider the average net capacity of the unit based on the emissions levels of the power plant as it show in equation 15:

$$P_{NET} = \alpha NOx^4 - \beta NOx^3 + \phi NOx^2 - \theta NOx + \omega \tag{15}$$

Where: α, β, ϕ, θ and ω are real constans values determined from data colection.

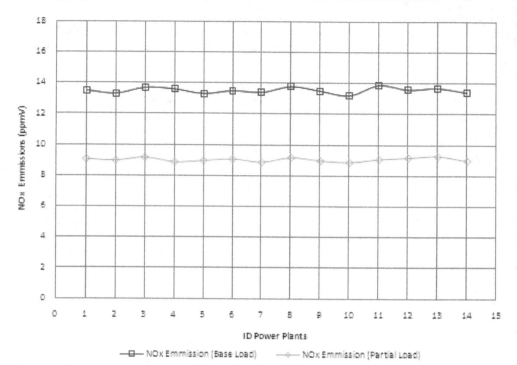

Fig. 17. Reference model of the variation rate in term of emissions for a power plant unit at base load or partial load (75 percent).

5. Conclusions

The goal of this work is to present the fundamentals for determining, through evaluation, a reference model for average net capacity and unit heat consumption as well as pollutant emissions level; furthermore the noise level in fourteen turbo gas power plants.

The model will serve as a reference in terms of Net Power and NOx emissions in the net effects of air quality due to this means of power generation. It will be possible to evaluate, during life history of the unit, the impact through the time due to natural degradation of equipment on the environment. This will be presented as a lower efficiency of the system and higher emissions and noise probably.

The noise levels have been determined for normal operation of unit to establish allowable limits to assure population's health and in the surroundings and personnel. Similarly, the NOx emissions and average net capacity model of reference will be a tool for comparison purposes in future operation.

On the other hand, as a main result of this analysis, it was verified that the unit observes all national and international regulations as well as all requirements previously established between costumer and manufacturer.

Finally, as a result of this study, it is guaranteed to the costumer that this plant will operate correctly satisfying technical requirements, according to unit design and local needs of generation in terms of efficiency and emissions.

6. Acknowledgment

The authors wish to acknowledge the support of the Mexican Federal Electricity Commission (Comisión Federal de Electricidad, CFE) to carry out this study.

7. References

[NOM-081-ECOL-1994 *"Establishment of the maximum permissible limits of noise emissions from fixed sources and its measurement method"*, in Spanish

Andrés A. Amell, César A. Bedoya, Bladimir Suárez, Efectos del cambio de composición química del gas natural sobre el comportamiento de turbinas a gas: Una aproximación al caso colombiano, *Revista Energética*, Vol. 35, 2006.

ASME PTC-22-1997., *"Performance Test Code on Gas Turbine"*, Performance Test Codes, ASME, 1997

Bengt Gudmundsson, Jenny Larfeldt, Stable dry low emission combustion with fuel flexibility, *7° Fórum de Turbomáquinas Petrobras*, Rio de Janeiro – 24 a 26 de novembro de 2009.

Christian Arnold, Hubert Skiba, Alexander Walke, Gas Market Liberalization in Germany and the Possible Consequences for Gas Turbine Operation, *OMMI* Vol. 2 Issue 3, December 2003.

D J Abbott, The impact of fuel gas composition on gas turbine operation: Future challenges, *British – French Flame Days*, Lille 9 / 10 March 2009.

G. Peureux, S. Carpentier, G. Lartigue, NOx emissions prediction for natural gas engines with fuel quality variations, *Proceedings of the European Combustion Meeting* 2009.

HQ-CTP904 - Commissioning Manual. *"Emissions Test Procedure"*, June, 2009.

K S Chana, K J Syed, M I Wedlock, R W Copplestone, M S Cook and G Bulat, Novel unsteady temperature/ heat transfer instrumentation and measurements in the presence of combustor instabilities, *ASME Turbo Expo, 6-9 June 2005*, Reno-Tahoe, Nevada, USA.

Lars O. Nord and Helmer G. Andersen, Influence of variations in the natural gas properties on the combustion process in terms of emissions and pulsations for a heavy-duty gas turbine, *Proceedings of International Joint Power Generation Conference*, June 16-19, 2003, Atlanta, Georgia, USA

NOM-085-ECOL-2003 *"Atmospheric pollution in stationary sources, for fixed sources that use solid fuels fossil, liquid or gaseous or anyone of its combinations that it establishes the maximum permissible levels of emission to the atmosphere of smoke and suspended particles"*, in Spanish.

Perspective of investment of the Mexican electrical sector, *COPARMEX-CFE*, June 2007

Report *"Distribution of industrial pollution by source of generation in Mexico"* RECT (Registro de Emisiones y Transferencias de Contaminantes, in Spanish), November 2007.

Robin McMillan, Peter Martin, Richard Noden, Mike Welch, Gas Fuel Flexibility in a Dry Low Emissions Combustion System, *Demag Delaval Industrial Turbomachinery Ltd.*, UK, February 2004.

S. Zaheer Akhtar. Gas Quality: Gas Turbine Boom Places Premium on Ensuring High-Quality Fuel Gas, *Power-Gen* Vol. 105, Issue 10, October 2001.

Secretary of the Environment and Natural Resources ("SEMARNAT" by its initials in Spanish), www.semarnat.gob.mx

Teresa Hansen, Meeting Gas Quality Challenges, *Power-Gen* Vol. 112, Issue 5, May 2008.

U.S. Environmental Protection Agency, http://www.epa.gov/

10

Modelling a SOFC Power Unit Using Natural Gas Fed Directly

Nguyen Duc Tuyen and Goro Fujita
Shibaura Institute of Technology
Japan

1. Introduction

1.1 Literature review and objective of this chapter

Completed models covering dynamic characteristics of those types of DGs are not openly available. The necessary task is to study these dynamic models based on the literature and any available operational data on DGs. Simulation of various types of DGs in a suitable software environment is the key step in analyzing the dynamic characteristics of DGs and designing the control strategies. In fact, computer simulation plays a vital role in the design and analysis of power system. Designing power systems without computer simulation is extremely laborious, time consuming, error-prone and expensive. Especially, in the new research field as DGs, computer simulation in an industrial environment with regard to the time in shortening the overall design process as compared to assembling and testing the components in the laboratory and deciding on the optimum values for components and controller parameters.

Among many types of FC, high-temperature fuel cells such as the solid oxide fuel cells (SOFC) have the potential for centralized power generation as well as combined heat and power. This chapter employs the SOFC model method. Especially, in this simulation, the rate of temperature change and load following ability will be included. In short, it consists of 3 main mathematical models, namely, the electrochemical model, the heat balance model and the power conditioning unit model.

The electrochemical model is to calculate output voltage, to regulate the fuel and air streams and to represent the ability to follow the load chance of SOFC.

The heat balance model is to calculate thermal energy inside SOFC stack as well as operating temperature. The heat exchangers are included in this simulation to represent the practical application when using to increase the temperature of input air and methane. Because of the high SOFC operating temperature, if there have no preheat, manufacture still have to set up a preheat system using electricity from other source or a small part of SOFC output power to elevate input species temperature to prevent thermal shock which can damage materials. Therefore, using heat generated inside SOFC stack for preheat can increase overall efficiency.

Finally, there are two important results that this chapter points out. The first is load following ability of SOFC power unit. The fuel cell control is achieved by adjusting the input volume of gas and air and controlling real power output. The two control loops which are in SOFC itself and in the DC-DC inverter make the SOFC power unit strongly following flexible change of load. And the second is heat balance inside fuel cell system with HX included.

1.2 Different modelling approaches

Depending upon the application, different models are available in the open literature and there are large differences in the level of details in the models presented. This section presents a review of the work of selected authors relevant to the model developed in the present work. Research work of SOFCs modelling has been begun since 1980's and there have been a lot of models developed so far. Initial models were lumped mass models and there were a lot of uncertainties in the results due to lack of experimental data as well as mature approaches. Increasing experimental research during early 1990's focused many such issues and established many empirical relationships to accurately predict the performance of SOFC. Also, due to the increase in calculation capabilities, it was possible to create more detailed models. During the late 1990's, several projects were initiated for detailed single SOFC modelling. Today modelling research is pursued in both detailed single SOFC modelling and system level stack modelling. Nevertheless, lumped models still continue to attract the attention of researchers due to their simplicity and small calculation time. Lumped models are considered over detailed models when it comes to predict accurately the FCs overall thermodynamic and electrical performance. A large amount of experimental data and mathematical relations exists for components such as air and fuel compressors, heat exchangers, thus these components can be modeled fairly accurately upon the lumped approach. Accordingly, lumped models are also easier to adjust to experimental data. The disadvantage of lumped SOFC models is that they can only account for mean values of the parameters and more detailed investigation of the cell is needed to check for undesirable effects such as thermal cracking, coking or exceeding temperature limits locally. This problem may be partly solved by using a detailed model to test the validity of the results after using a lumped model for system calculations. Obviously, implementing a detailed SOFC model in the system model gives the most accurate results.

References (Achenbach & Elmar, 1995), (Wang & Nehrir, 2007) provide a basic approach for fuel cell modelling suitable for distributed generation, however not discussed in details about SOFC. A SOFC model has been developed by various researchers in (Li & Chyu, 2003), (Ali Volkan Akkaya & Erdem, 2009), (N. Lu, 2006), (David L. Damm, 2005), (Mitsunori Iwata, 2000), (Tomoyuki Ota, 2003), (Xiongwen Zhang, 2007), (Takanobu Shimada, 2009), (S. Campanari, 2004) and (A.C. Burt, 2004) taking its thermodynamic effect into consideration which concentrated on the effects of temperature changes on the output voltage response. Heat balance is considered in specific model configuration more than in general and the detail calculation seems to be complex. Some did not consider the dynamics of the chemical species. Reference (Tadashi Gengo, 2007) points out empirical responses of real model considering temperature inside SOFC combine with output voltage, current and power. (Takanobu Shimada, 2009), (D. Sanchez, 2008) considers internal CH_4 reformer, fuel and air input temperature are increased. But there is not any research caring about using SOFC exhaust to take full advantages of high operating temperature. (M. Uzunoglu, 2006), (Caisheng Wang, 2007), (Caisheng Wang, 2005) take the double layer charging effect into account but not SOFC. Recently research (Takanobu Shimada, 2009), (P. Piroonlerkgul, 2009),

(Graham M. Goldin, 2009) approach this chapter ideal when taking HX into account but calculate heat balance in another way and for specific SOFC configuration. David investigated the transient behavior of a stand-alone SOFC caused by a load change in (Achenbach & Elmar, 1995), (Kourosh Sedghisigarchi, 2004), (J. Padulles, 2000). However, the built model is simple for evaluate the real response but these simplified models consider constant cell temperature. (M.Y. El-Sharkh, 2004) considers only the dynamic characteristic of Power Conditioning Unit system. (P.R. Pathapati, 2005) represents PEM dynamic model which does not consider concentration loss. (S.H.Chan, 2002) deals with HX model. A physically based model for tubular SOFC was developed in (Caisheng Wang, 2007).

A transient dynamic model of SOFC will be proposed in this chapter. Electrochemical and thermal simulations of a SOFC reported in all reference will be used to identify the key parameters of this SOFC system from a single cell to a N_0 single fuel cell connected in series. The cell's terminal voltage during a load change was discussed. Overall heat balance inside SOFC power unit effecting on operating temperature afterward on output voltage will be pointed out.

1.3 Fuel cell

Fuel cells will be important components of distribution system due to their high efficiency and low environmental pollution. Generally, efficiency of the fuel cells ranges from 40-60% can be improved to 80-90% in co-generation applications. The waste heat produced by the lower temperature cells is undesirable since it cannot be used for any application and thus limits the efficiency of the system. The higher temperature fuel cells have higher efficiency since the heat produced can be used for heating purposes. Due to an electric-chemical process of power generation, there is no noise develop usually in mechanical members of conventional generator. All of these features will without any doubt lead to their wide application in the power industry in the near future. Several types of fuel cells have been reported in the literatures: phosphor acid fuel cell (PAFC), solid oxide fuel cell (SOFC), molten carbonate fuel cell (MCFC) and proton exchange membrane fuel cell (PEMFC). The PAFC has been commercially used in hospitals, nursing homes, utility power plants, etc. The SOFC can be used in large-power applications such as central electricity generation station. SOFC has the highest potential in large power application.

1.4 Fuel for fuel cells

Each of these FC types differs in the electrolyte and fuel used, operating temperature and pressure, construction materials, power density and efficiency.

The most important component of a FC is the fuel processor and the reformer since hydrogen is not readily available. Fossil fuels such as gasoline, natural gas and coal gases need to be processed and reformed to obtain enriched hydrogen. Natural gas is the most easily available fuel source. Bio-fuels can also be used as a source to obtain hydrogen. Biological methods such as photosynthesis and fermentation can be used to produce hydrogen. Though there are different methods to produce hydrogen, a proper and feasible method which can be commercialized is not yet available.

Storage of hydrogen is an important aspect of the FC systems because the fuel has to be readily available for continuous supply of electric power. Sometimes, electrical energy is used to divide water into hydrogen and oxygen with the help of electrolysers during times of high supply and low demand. FCs have to be compact and portable for mobile applications; hence storage of hydrogen is essential for such applications. Hydrogen needs to be handled with great care because it is a highly volatile and flammable gas. It has a high leak rate due to which the gas tends to escape through small orifices, faster than the other gases. Hence storage of hydrogen plays a key role in the FC systems.

A fuel processor converts the primary fuel source (hydrocarbons) into the fuel gas (hydrogen) required by the FC stack. The processor uses a catalytic reaction to break the fuel into hydrogen and separate it from the carbon based gases. Each of the FC types has specific fuel requirements. Natural gas and petroleum liquids contain sulphur compounds and have to be desulphurized before they can be used as a fuel. The anode catalysts are intolerant of sulphur and it must be removed before it degrades catalyst performance. There is a risk of carbon formation in fuel cell systems which can be reduced by carrying out pre-reforming of the fuel gas before it is fed to the reformer reactor. Carbon monoxide can be used as a fuel for SOFC and MCFC because it can be internally converted to hydrogen whereas the PEMFC should be completely free from it. CO has high affinity for anode catalyst (especially platinum) and it prevents the flow of fuel in the PEMFC. Ammonia is a poison for all the FC types due to its adverse effects on the cell life except for SOFC, where it can be internally reformed.

Lower-temperature FCs require an external reformer to obtain the hydrogen rich fuel, thus increasing the cost and thereby reducing the efficiency. Higher temperature FCs do not require an external reformer; its high temperature allows direct conversion of natural gas to hydrogen. High temperature requires stringent materials which increases the cost of the fuel cells. Hence, researchers are working to combine the benefits of the PEMFC and the PAFC to obtain intermediate temperature cells, often referred to as high temperature PEM.

2. Solid oxide fuel cell

The SOFC technology, dates from Walther Nernst, who around 1890 discovered that stabilized zirconia is an isolator at room temperature, but turns into an ionic conductor between $600 - 1000^{\circ}C$ and an electronic and ionic conductor around $1500^{\circ}C$. The first SOFC based on zirconia was introduced by Baur and Preis in 1937. Since then, research on SOFC has been steadily increasing until today.

2.1 Fundamentals

There are different types of FCs that have been mentioned above and are currently in use and development. Among them, SOFCs that works in the simplest structure have grown in recognition as a viable high temperature FC technology. SOFC can be improved to create a hydrogen fuel with heat inevitably occur in the cell, the power structure is a device that requires no external reformer, and without the need for transformer and reactor. The SOFC has a few typical advantages compared to other FCs.

1. Higher efficiency compared to other FCs
2. Easy to handle with simple structure which is composed of all solid
3. The reforming system is simple
4. Carbon monoxide (CO) can be used as a fuel

5. No precious metals as catalysts (platinum: Pt, etc.)

6. Structure for carbon dioxide recovery

In addition, while maintaining high efficiency, to lower the temperature of the SOFC is driving development issues. Because $1000^{\circ}C$ high operating temperature of the material will degrade cells, reduce the choice of construction materials. Operating temperature of $700 - 1000^{\circ}C$ from the conventional $500 - 800^{\circ}C$, these issues are resolved smaller, lower cost, can improve endurance. High-temperature operation removes the need for a precious-metal catalyst, thereby reducing the cost. It also allows SOFCs to reform fuels internally, which enables the use of a variety of fuels and reduces the cost associated with adding a reformer to the system. Although a SOFC produces electricity, it only produces DC power and utilizes only processed fuel. Therefore, a SOFC based power generation system requires the integration of many other components beyond the SOFC stack itself. Moreover, to recover the high quality waste heat from the SOFC stack, an efficient integration of co-generation or bottoming system with the FC section is crucial for a SOFC based power generation plant. Since the balance of plant will directly impact the overall system efficiency and may cost more than the SOFC stack itself, it is obvious that the design of a SOFC power generation system involves more than the optimization of the SOFC unit with respect to efficiency or economics. It also involves balance of plant studies. With SOFC materials and stacks approaching a commercialization stage, there is a need to explore various process designs to obtain optimal efficiency and economics based on specific applications and fuel availability.

2.2 SOFC based power generation systems

As other types of FCs, a SOFC produces only DC power and requires processed fuel. It also produces high quality heat due to its high operating temperature. Beyond the SOFC stack itself, a typical SOFC power system basically includes: a reformer to start the hydrogen production process, a fuel conditioner to clean up the pollutants that could otherwise poison the fuel cell elements, a power conditioner to convert direct current from the fuel cell to the appropriate voltage range and current type depending on the application, and a cogeneration or bottoming cycle to utilize the rejected heat to achieve high system efficiency. The system also requires the most common balance of plant equipments such as heat exchangers, air blower and fuel compressors, controls systems, and safety systems. Fig.1 illustrates fundamental parts in a FC power unit.

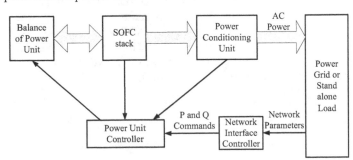

Fig. 1. Structure of SOFC power generation unit

2.3 Fuel processing

Fuel processing converts a commercially available fuel to a fuel gas suitable for the SOFC anode reaction. Typical fuel processing steps include:

- Desulphurization, where a catalyst is used to remove sulphur contaminants in the fuel. Sulphur compounds are noxious, and they can also bind catalysts used in later stages of fuel reformation poisoning the catalyst.
- Reformation, where the fuel is mixed with steam and then passed over a catalyst to break it down into hydrogen, as well as carbon dioxide and carbon monoxide.
- Shift conversion, where the carbon monoxide reacts with steam over a catalyst to produce more hydrogen and carbon dioxide.

However, high operating temperature SOFCs can accommodate internal reforming by means of a CO-tolerant nickel catalyst, so they can operate on natural gas with minimum pre-processing of the fuel. This will not only reduce the capital cost of the SOFC system, but also can be beneficial to system efficiency because there is an effective transfer of heat from the exothermic cell reaction to satisfy the endothermic reforming reaction.

Hydrogen sulfide, hydrogen chloride and ammonia are impurities typically found in coal gas. Some of these substances maybe are harmful to the performance of SOFCs. Therefore, a SOFC system will require fuel cleanup equipment such as desulfurizer depending on the raw fuel components.

2.4 Rejected heat utilization

At $1000C$ operating temperature, SOFCs produce a tremendous amount of waste heat while generating electricity. In order to obtain the highest possible system efficiency, the heat must be recovered by producing hot water, steam, or additional electricity. In a large SOFC power system (>100MW), production of electricity via a steam turbine bottoming cycle is maybe advantageous.

2.5 Power conditioning unit

While used as a power generator, FCs usually are connected to the load or distribution system via Power Conditioning Unit basically including DC-DC converter and DC-AC inverter. Therefore, low cost and high efficiency inverters are required together with acting controllers for fast tracking of real and reactive power demands. The inverter serves as the interface between the SOFC and the power distribution system. It is controlled in order to provide real and reactive power set point tracking and to adjust the power factor as well as frequency. Transient response control equipment may also be included. The efficiency of the power conversion is typically on the order of 94 to 98%.

2.6 Electrochemistry of SOFC

Fig.2 shows the processes taking place in a SOFC with hydrogen.

The SOFC fundamentally consists of two porous electrodes (anode and cathode) separated by a ceramic electrolyte in the middle, and flow channels for fuel and air delivery and collection.

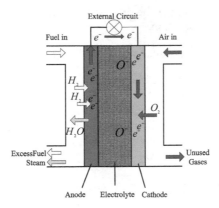

Fig. 2. Reaction process inside SOFC

Air flows along the cathode. When an oxygen molecule inside air flow contacts the cathode/electrolyte interface, it acquires 4 electrons from the cathode and splits into 2 oxygen ions. The reaction occurring at the cathode is:

$$O_2 + 4e^- \rightarrow 2O^{2-} \tag{1}$$

The oxygen ions diffuse into the electrolyte material and migrate to the other side of the cell where they encounter the anode. These oxygen ions travel through the porous electrolyte and react with H2 (fuel) to produce water and most importantly, electrons at the anode and the following reaction may occur at the anode:

$$2H_2 + 2O^{2-} \rightarrow H_2O + 4e^- \tag{2}$$

The electrons transport through the anode to the external circuit and back to the cathode, thus an electro motive force (EMF) is generated between two electrodes. The two electrodes can be connected via an external circuit and an electrical current can be generated. At the same time, we get heat when the reaction occurs. The overall reactions are:

$$H_2 + \frac{1}{2}O_2 \rightarrow H_2O \tag{3}$$

In cases using methane as a fuel, the CH4 reacts with O2 via internal reformer which will be discuss later that produces H2O and CO2. The final reaction equation is:

$$CH_4 + 2O_2 \rightarrow 2H_2O + CO_2 \tag{4}$$

2.7 Fuel cell voltage and nernst equation

Before we begin to look at how the electromotive force (EMF) and thus work is produced in a FC, it is necessary to understand some basic thermodynamic concepts. The Gibbs free energy is the energy required for a system at a constant temperature with a negligible volume, minus any energy transferred to the environment due to heat flux. Gibbs free energy is the energy available to do external work which involves moving electrons around an external circuit. In FCs, change in Gibbs free energy of formation (ΔG) is considered, as this change is responsible for the energy released. This change is the difference between the free energy of the products and the reactants, as shown in equation.

$$\Delta G = \Delta G_{products} - \Delta G_{reactants} \tag{5}$$

Consider the following thermodynamic identity for a reversible process when there is no shaft work extracted and the system is restricted to do only expansion work: $dG = VdP - SdT$, and if the process is isothermal, the above equation reduces to: $dG = VdP$. Using the ideal gas equation, $V = nRT$, we have $dG = nRTdP/P$. Integrating this equation from state 1 to state 2, we get $G_1 - G_2 = nRTln\frac{P_2}{P_1}$. If the state 1 is replaced with some standard reference state, with Gibbs free energy G_0 and standard pressure P_0, the Gibbs free energy per unit mole at any state 'i' is given by,

$$g_i = g_0 + RTln\frac{P_i}{P_0} \tag{6}$$

Consider that the following chemical reaction takes place at constant pressure and temperature, $aA + bB \leftrightarrow mM + nN$ Where a, b, m and n are the stoichiometric coefficients of the reactants A and B and the products M and N, respectively. Now, Equation 6 takes the following form,

$$\Delta G = \Delta G_0 + RTln(\frac{P_M^m P_N^n}{P_A^a P_B^b}) \tag{7}$$

ΔG_0 is the standard Gibbs free energy change for the reaction ($\Delta G_0 = mg_M^0 + ng_N^0 - ag_A^0 - bg_B^0$ and g_i^0 are the standard Gibbs free energies of the constituents).

Equation 7 gives the Gibbs free energy change for the reaction. We are interested on how is that energy change is related to the work of the SOFC system performed. To find that relation, consider the following thermodynamic identity for a reversible process, $(dQ = TdS)$

$$dG = -\delta W + PdV + VdP - SdT \tag{8}$$

At constant temperature and pressure, the above equation can be written as,

$$dG = -\delta W + PdV \tag{9}$$

Since it is a non-expansion work, Equation (2.9) takes the form,

$$dG = -\delta W_e \tag{10}$$

Equation 10 means the change in Gibbs free energy of the reaction is equal to the maximum electrochemical work, W_e, that can be extracted when reactants A and B react to give products M and N under constant temperature and pressure conditions through a reversible reaction.

Now, we can focus on how the maximum electrochemical work relates to the EMF of the cell.

For the SOFC, n_e (n_e=8 with Equation 4(of Energy, 2004)) electrons pass through the external circuit for each CH_4 molecule used. In a lossless system, electrical work done is equal to the change in Gibbs free energy which has been proved previously. Further, electrical work done to move a charge of n_e F (to move n_e electrons) for a voltage of E is given by below equation.

$$\text{Electrical work done} = -n_e FE_{cell} \text{joules} \tag{11}$$

(1e charge 1.602×10^{-19}, therefore 1 mole CH_4 which is equivalent to n_e mole e (1mol has $N = 6.022 \times 10^{23}$ e) will charge $-n_e \times 6.022 \times 10^{23}, 1.602 \times 10^{-19} = -n_e \times 96485 = n_e \times F$)

The EMF produced due to half-cell reactions drives the electrons to move from the anode to the cathode. If ne mole of electrons move from anode to cathode per unit time and the EMF of the cell is E, the power extracted is simply EMF multiplied by the current,

$$W_e = n_e F E_{cell} \tag{12}$$

where F is the total charge of 1 mole of electrons, known as Faraday's constant. Now if we look at the integral form of Equation 7), 10 combined with 12, we get,

$$\Delta G = n_e F E_{cell} \tag{13}$$

Applying Equation 7 to Equation 13 we get what is known as Nernst equation,

$$E_{cell} = \frac{\Delta G_0}{n_e F} + \frac{RT}{n_e F} ln\left(\frac{P_M^m P_N^n}{P_A^a P_B^b}\right) \tag{14}$$

For the reaction occurring in an SOFC with methane in Equation 4:

$$E_{cell} = \frac{\Delta G_0}{n_e F} + \frac{RT}{n_e F} ln\left(\frac{P_{CH_4} P_{O_2}^2}{P_{H_2O}^2 P_{CO_2}}\right) = E_0 + \frac{RT}{8F} ln\left(\frac{P_{CH_4} P_{O_2}^2}{P_{H_2O}^2 P_{CO_2}}\right) \tag{15}$$

With E_0: Ideal Voltage at Standard Pressure and: $\Delta g_0 = 2g_{H_2O}^0 + g_{CO_2}^0 - 2g_{O_2}^0 - g_{CH_4}^0 = -980(kJ.mol^{-1})$(at standard pressure 1 atm and temperature 298 K). (Δg_0 changes with reaction like Equation 4), it is fairly constant with temperature)

Therefore:

$$E_0 = \frac{980.000}{8 \times 96485} = 1.27[V] \tag{16}$$

Actually E_0 depends on temperature:

$$E_0^T = E_0 - k_E(T - 298) \tag{17}$$

This maximum theoretical voltage, E, is also known as "Open Circuit Voltage" (OCV) and can be measured when there is no current in the circuit. Also, it can be observed, that to get the maximum OCV, a high concentration of reactants is required.

2.8 Voltage loss

When the FC is under load (a current is flowing), the voltage supplied at the electrodes will be different from the E_{cell} calculated from Equation 15. The dependency of these losses on temperature, current density and species concentrations mainly determine the characteristics of a FC. The output voltage is therefore lower than the circuit voltage when the FC is operated. Three main mechanisms of voltage losses exist: activation/polarization loss (η_{act}), Ohmic loss (η_{ohmic}), and concentration/diffusion loss (η_{con}).

A typical curve of the cell electrical voltage against current density is shown in Figure 3. It can be seen that there exists a linear region where the voltage drop is linearly related with the current density due to the Ohmic contact. Beyond this region the change in output voltage varies rapidly. At very high current density, the voltage drops significantly because of the gas exchange efficiency (η_{con}). At low current level, the Ohmic loss becomes less significant, the increase in output voltage is mainly due to the activity of the chemicals (η_{act}).

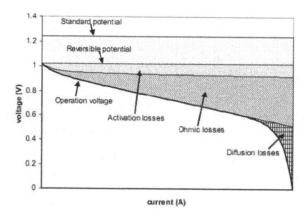

Fig. 3. FuelCell Output Voltage (of Energy, 2004)

The output voltage of a cell, V_{cell} can, therefore, be written as:

$$V_{cell} = E_{cell} - \eta_{act} - \eta_{ohmic} - \eta_{con} \tag{18}$$

The output voltage of the FC stack including N_0 individual single FC connected in series can be obtained as:

$$V = N_0[E_{cell} - \eta_{act} - \eta_{ohmic} - \eta_{con}] \tag{19}$$

To calculate the FC output voltage, the above three voltage drops should be calculated.

2.8.1 Activation loss

Action voltage loss is caused by an activation energy barrier that must be overcome before the chemical reaction occurs. At open circuit, no outer current is flowing. However, reactions are still taking place, but at equal rates in both directions. Just regarding the current which flows into one of the directions, we find the "exchange current density". In order to achieve an outer current higher than this, an extra potential is required to achieve the desired reaction rate, called activation voltage. The voltage drop is increasing fast at low reaction rates and is from a certain level almost constant. Activation is the dominant source of loss for low-temperature FCs, while their influence is smaller for SOFCs.

Butler-Volmer equation is normally used to calculate the activation voltage drop. To avoid the ambiguity of simplified model, such as the Tafel equation or a linear potential-current relation, used under different operating conditions, the following general Butler-Volmer equation is used to calculate the respective overpotential of anode and cathode:

$$i = i_{0,k} \left\{ e^{\beta \frac{n_e F \eta_{act,k}}{RT}} - e^{-(1-\beta)\frac{n_e F \eta_{act,k}}{RT}} \right\} \tag{20}$$

- β: transfer coefficient. (is considered to be the fraction of the change in polarization that leads to the change in reaction rate constant; its value is usually 0.5 in the context of a FC).
- i_0: the 'apparent' current exchange density
- k: is anode (a) or cathode (c)

Under high activation polarization, the first exponential term in Equation 19 will be much less than unity and the exponential terms can be excluded from the equation. Rearranging the simplified equation yields $\eta_{act} = \frac{RT}{\beta n_e F} ln \frac{i}{i_0}$ which is the well-known Tafel equation. This equation will yield an unreasonable value for $\eta_a ct$ when $i = 0$ (in the case of load is be disconnected to SOFC because of faults). Therefore, we will use the Butler-Volmer equation. Hence,

$$i = 2i_{0,k}sinh(\frac{n_e F \eta_{act,k}}{RT}) \rightarrow \eta_{act,k} = \frac{2RT}{n_e F}sinh^{-1}(\frac{i}{2i_{0,k}}) \tag{21}$$

From (S.H. Chan, 2001),

$$i_{0,a} = 5300[A/m^2] = 5300.1000/10000 = 530[mA/cm^2] \tag{22}$$

$$i_{0,c} = 2000[A/m^2] = 2000.1000/10000 = 200[mA/cm^2] \tag{23}$$

Actually, the temperature effects on exchange current density. However in this simulation, this influence is so small that can be neglected. Hence,

$$\eta_{act} = \eta_{act,a} + \eta_{act,c} = \frac{2RT}{n_e F}[sinh^{-1}(\frac{i}{2i_{0,a}}) + sinh^{-1}(\frac{i}{2i_{0,c}})] \tag{24}$$

The equivalent activation resistance can then be defined as:

$$R_{act} = \frac{\eta_{act}}{i} = \frac{2RT}{i n_e F}[sinh^{-1}(\frac{i}{2i_{0,a}}) + sinh^{-1}(\frac{i}{2i_{0,c}})] \tag{25}$$

Compared with Anode, the Cathode exhibits higher activation overpotential, which is due to the poor "apparent" exchange current density at the electrode/electrolyte (LSM-YSZ/YSZ) interface. Since the Cathode exchange current density directly affects the electrochemical reaction rate at the Cathode, it can be understood that the low electrochemical reaction rate in the Cathode lead to high cathode activation polarization in the SOFC.

According to Equation 24, the activation voltage drop will be zero when load current is zero. The Ohmic and concentration voltage drops (will be discussed) are also zero when the fuel cell is not loaded ($i = 0$). However, even the open-circuit voltage of an SOFC is known to be less than the theoretical value given by Equation 25). Therefore, a constant and a temperature-dependent term can also be added to Equation (2.24) for activation voltage drop computation of SOFC as follows (P.R. Pathapati, 2005):

$$\eta_{act}^T = \xi_1 + \xi_2 T + iR_{act} = \eta_{act,1} + \eta_{act,2} \tag{26}$$

where $\eta_{act,1} = \xi_1 + \xi_2 T$ is the part of activation drop affected only by the FC internal temperature, while $\eta_{act,2} = iR_{act}$ is both current and temperature dependent.

2.8.2 Ohmic overpotential

Ohmic overpotential, which contributes by the electrolyte, electrodes and interconnector of the FC, occurs because of the resistance to the flow of ions in the ionic conductors and the resistance to electrons through the electronic conductors. At a given temperature and geometry, the voltage loss is proportional to the current. Since these resistances obey Ohm's law, the overall Ohmic overpotential can be written as:

$$\eta_{Ohmic} = iR_{ohm} \tag{27}$$

The resistances of these FC components are determined by the resistivity of the materials used and their respective thickness. The results show that the resistances of the cathode, electrolyte and interconnector decrease with increase in temperature. By contrast, the anode resistance displays the opposite trend. The resistance of each material used in the SOFC components can be calculated from its respective resistivity, which is a function of temperature. The electrical resistance R_{ohm} is calculated simply as the sum of the anode R_a, electrolyte R_e, cathode R_c and interconnector resistance R_i:

$$R_{ohm} = R_a + R_e + R_c + R_i \tag{28}$$

However, the main contribution to the Ohmic polarization is from the transport resistance of O^{2-} in the electrolyte. The resistance of electrolyte strongly depends on the temperature, and its effect cannot be ignored. The dependence of electrolyte resistance on temperature is given by the following equation:

$$R_e = R_{e0}e^{10100(\frac{1}{T} - \frac{1}{1273})} \tag{29}$$

The resistances of other parts are assumed to be constant because of the weak dependence on temperature and their contributions to the total voltage drop are small (Susumu Nagata, 2001). Hence,

$$R_{ohm} = R_{e0}e^{10100(\frac{1}{T} - \frac{1}{1273})} + R_a + R_c + R_i \tag{30}$$

According to (Takanobu Shimada, 2009),

$$\begin{aligned} R_{ohm} &= R_{e0}e^{10100(\frac{1}{T} - \frac{1}{1273})} + R_a + R_c + R_i \\ &= \frac{0.00745}{0.3} \times e^{10100(\frac{1}{T} - \frac{1}{1273})} + (0.01003 + 0.0184) \times 0.3 + \frac{0.00022}{0.3} \\ &= 0.0248 \times e^{10100(\frac{1}{T} - \frac{1}{1273})} + 0.0093 [\Omega/cm^2] \end{aligned} \tag{31}$$

2.8.3 Concentration overpotential

Reactants must flow through the porous electrodes to the TPB, and products must flow into the other direction, driven by diffusion. During the reaction process, concentration gradients can be formed due to mass diffusion from the flow channels to the reaction sites (catalyst surfaces). The effective partial pressures of hydrogen and oxygen at the reaction site are less than those in the electrode channels, while the effective partial pressure of water at the reaction site is higher than that in the anode channel. Thus, the calculated potential will be lower and the difference is called diffusion or concentration losses. At high-current densities, slow transportation of reactants (products) to (from) the reaction site is the main reason for the concentration voltage drop. Any water film covering the catalyst surfaces at the anode and cathode can be another contributor to this voltage drop. The voltage drop increases with increasing current against an asymptotic maximum current. At this point, the concentration of one of the reactants at the TPB is zero and no further current increase is possible. This equation is entirely empirical which has become more favored lately after many research finding out accurate equations, and yields an equation that fits the results very well. It provided the constants $m = 3.10^{-5}\,[V]$ and $n = 8.10^{-3}\,[mA^{-1}.cm^2]$ are chosen properly.

$$\eta_{con} = me^{ni_{fc}} \tag{32}$$

The equivalent resistance for the concentration voltage drop can be calculated as:

$$R_{con} = \frac{\eta_{con}}{i_{fc}} \tag{33}$$

2.9 Double-layer charging effect

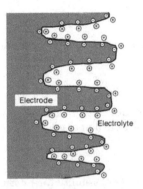

Fig. 4. The charge double layer at the surface of a fuel cell electrode (James Larminie, 2003)

In an SOFC, the two electrodes are separated by the electrolyte (Figure 4), and two boundary layers are formed, e.g., anode-electrolyte layer and electrolyte-cathode layer. These layers can be charged by polarization effect, known as electrochemical double-layer charging effect, during normal fuel cell operation. The layers can store electrical energy and behave like a super-capacitor. The model for SOFC considering this effect can be described by the equivalent circuit shown in Figure 5 (Caisheng Wang, 2005).

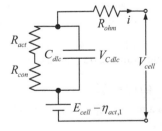

Fig. 5. Equivalent electrical circuit of the double-layer charging effect inside an SOFC

In the above circuit, R_{ohm}, R_{act} and R_{con} are the equivalent resistances of Ohmic voltage drop, activation, and concentration voltage drops, which can be calculated according to Equation

31, 25, and 33, respectively. C_{dlc} is the equivalent capacitor due to the double-layer charging effect. The capacitance of a capacitor is given by this formula:

$$C_{dlc} = \varepsilon \frac{A}{d} \tag{34}$$

where ε is the electrical permittivity, A is the surface area, and d is the separation of the plates. In this case, A is the real surface area of the electrode, which is several thousand times greater than its length×width since the electrodes of a SOFC fuel cell are porous. Also d, the separation, is very small, typically only a few nanometers. The result is that, in some fuel cells, the capacitance will be very large (can be in the order of several Farads) which is high in terms of capacitance values. (In electrical circuits, a 1 μF capacitor is on the large size of average). The voltage across C_{dlc} is :

$$V_C = (i - C_{dlc} \frac{dV_C}{dt})(R_{act} + R_{con}) \tag{35}$$

The double-layer charging effect is integrated into the modelling, by using V_C instead of $\eta_{act,2}$ and η_{con} to calculate V_{cell} . The fuel cell output voltage now turns out to be:

$$V_{cell} = N_0[E_{cell} - V_C - \eta_{ohmic} - \eta_{act,1}] \tag{36}$$

Recent approaches show that cathode activation and Ohmic overpotentials are responsible for the major losses in the SOFC over normal operating range.

2.10 Reforming

A great advantage of the SOFC is its possibility for internal reforming of hydrocarbon fuel. Sulphur-free natural gas (mainly a mixture of the alkanes methane, ethane and propane), which is technically available today, may be used as fuel. Due to the high temperature and the existence of nickel as a catalyst at the anode, the fuel cell reforms the alkanes to hydrogen and carbon monoxide internally through the steam reforming reaction:

$$C_n H_{2n+2} + n H_2O \Leftrightarrow (2n + 1)H_2 + nCO$$

The equilibrium of this reaction is at the right hand side for elevated temperatures. As the reforming reaction is strongly endothermic, it severely decreases the temperature where it takes place in the fuel cell and therewith the local current density.

The model is considering reforming the internal reformer, but rather a reaction when hydrogen and oxygen that is generated from methane. Equation 37 shows the internal reforming reaction formula

$$CH_4 + H_2O \rightarrow 3H_2 + CO \tag{37}$$

1 [mol] CH_4 generates 3 [mol] H_2 and 1 [mol]CO. These products react with O_2. The following equations:

$$3H_2 + 3/2O_2 \rightarrow 3H_2O \tag{38}$$

$$CO + 1/2O_2 \rightarrow CO_2 \tag{39}$$

These three equations match the combustion reaction of methane shown in this equation 40. $CH_4 + H_2O + 2O_2 \rightarrow 3H_2O + CO_2$ Or

$$CH_4 + 2O_2 \rightarrow 2H_2O + CO_2 \tag{40}$$

The Equation 40 is equivalent to Equation 4. In short, internal reforming model is used to model the reaction of methane combustion.

3. Heat exchangers

Heat exchangers are used extensively in the energy and process industry. In power cycles they are called recuperators and their use is to recover heat from exhaust streams for preheating the process streams and therewith saving part of the fuel. SOFC systems in particular involve recuperation of heat due to the high gas inlet temperatures required and the high amount of heat in the exhaust. In fact the inlet temperatures of air and fuel have to be increased somehow to get out of heat shock with the reactants. The reason here is that it is cannot to fed the inlet species with low temperature such as ambient temperature while the stack temperature is quite high. By any kind of means, SOFC power unit has to increase the inlet species temperature, for example, using several percentage of output SOFC power. In this simple SOFC power unit model, waste heat recovery used for preheating the fuel and air directly, the system efficiency therefore can be improved.

3.1 Configuration

There are different flow configurations, depending on the application. Analogous to the fuel cell, flow configurations may be co-flow, counter-flow or cross-flow. Exegetically, counter-flow is most efficient, because the cold fluid outlet may closely approach the hot fluid inlet temperature if the flow rates and HX surface are suitably chosen. A co-flow configuration may be more effective for HXs with a huge temperature difference between hot and cold fluid and only small temperature changes. In this model, both counter flow and parallel flow types of HX are selected as pre-heaters for comparison.

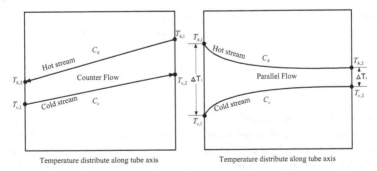

Fig. 6. Temperature distribution of flows in HX

The Simulink Model has been designed to assume that the inlet temperatures of air and fuel are equal to ambient temperature, i.e. $T_1 = 298[K]$.

To determine the outlet temperatures of the pre-heats, which vary with the inlet conditions, the heat capacity rates of the cold and hot gas streams are calculated:

$$\begin{cases} C_c = \sum q_c C_{p,c} \\ C_c = \sum q_c C_{p,c} \end{cases} \left[\frac{mole}{s} \cdot \frac{J}{mole.K} = W/K \right] \tag{41}$$

From Equation 4: $CH_4 + 2O_2 = 2H_2O + CO_2$

Species	Heat Capacity C $(J/mole.K)$
CH_4	75.264
CO_2	57.112
H_2O	75.312
O_2	35.84
N_2	33.964
Air $(21\%O_2 + 78\%N_2)$	34.018

Table 1. Heat capacity of SOFC species

- HX1

$$C_c[\text{Air}] = q_{O_2}^{in} C_{O_2} + (\frac{78}{21}) q_{O_2}^{in} C_{N_2} \tag{42}$$

$$\begin{aligned} C_h[CH_4 + N_2 + CO_2 + H_2O] &= \\ &= \left(\frac{1-U_{opt}}{U_{opt}} \right) q_{CH_4}^{in} + (\frac{78}{21}) q_{O_2}^{in} C_{N_2} + \left(\frac{1-U_{opt}}{U_{opt}} \right) q_{O_2}^{in} C_{O_2} + q_{H_2O}^{out} C_{H_2O} + q_{CO_2}^{out} C_{CO_2} \end{aligned} \tag{43}$$

- HX2

$$C_c[CH_4] = q_{CH_4}^{in} C_{CH_4} \tag{44}$$

$$C_h[HX_2] = C_h[HX_1] \tag{45}$$

The effectiveness & number of transfer units method (ε-NTU methodology: ε=f[NTU,C_r])) is used to model the pre-heaters in CH_4 SOFC system, which makes use of two non-dimensional groups: the number of heat transfer units - NTU, and the effectiveness, ČÃ defined below.

By comparing two heat capacities of hot and cold streams, the lower and higher values are assigned as C_{min} and C_{max}, respectively. The ratio of heat capacity rates is then available. Thus,

$$C_r = C_{min}/C_{max} \tag{46}$$

Number of heat transfer unit,

$$NTU = UA/C_{min} \tag{47}$$

where, U is overall heat transfer coefficient [W/cm^2K], which is defined largely by the system and in many cases it proves to be insensitive to the operating conditions of the system. With our simulation, we take U to be a constant value and $U = 0.5[W/cm^2K]$ with high pressure gas. This is fairly reasonable in compact single-phase heat exchangers; and A-total heat transfer area.

The heat exchanger effectiveness is defined as the ratio of actual heat transfer rate, q, and the maximum possible heat transfer rate between the 2 streams, q_{max},

$$\varepsilon = q/q_{max} \tag{48}$$

Hence, the heat exchange rate between the hot and the cold gas stream is:

$$q = \varepsilon q_{max} \tag{49}$$

Where the theoretical maximum heat transfer is:

$$q_{max} = C_{min}(T_{h,1} - T_{c,1}) \tag{50}$$

With the value $(T_{h,1} - T_{c,1})$ is simply the temperature difference between the 2 inlet ports, i.e. the largest temperature difference between 2 streams and hence defines the ceiling value for the heat transfer rates between the 2 streams.

$$\varepsilon = \frac{C_h(T_{h,1} - T_{h,2})}{C_{min}(T_{h,1} - T_{c,1})} = \frac{C_h(T_{c,2} - T_{c,1})}{C_{min}(T_{h,1} - T_{c,1})} \tag{51}$$

And we have the ε-NTU relation by these equations:

- In case of counter flow:

$$\varepsilon = \frac{1 - e^{[-NTU(1-C_r)]}}{1 - C_r e^{[-NTU(1-C_r)]}} \tag{52}$$

- In case of parallel flow:

$$\varepsilon = \frac{1 - e^{[-NTU(1+C_r)]}}{1 + C_r} \tag{53}$$

Therefore, once effectiveness is calculated, based on the energy balance, the exit temperature of the hot and cold gas streams from the heat exchanger are:

$$T_{h,2} = T_{h,1} - q/C_h \tag{54}$$

$$T_{c,2} = T_{c,1} + q/C_c \tag{55}$$

3.2 Gas temperature of SOFC exhaust

Because relatively high amount of input gas flow not be used and they pass though SOFC without reaction. With the reactants, it can be considered their output temperate is operating temperature. However, non-reactants do not have the operating temperature when they come out from SOFC stack. Assuming that the SOFC configuration is as a Heat Exchanger for these gases, it is similarly to calculate their output temperature as in case of HX in the above section. Following figure represents the concept of this idea.

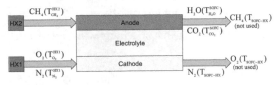

Fig. 7. Exhaust temperature calculation concept

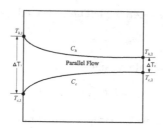

Fig. 8. Considering temperature change of SOFC exhaust as in HX

And SOFC is considered as HX to calculate non-reactants output temperature. Heat capacity rates of the cold and hot gas streams are calculated:

$$C_h[CO_2 + H_2O] = q_{CO_2}^{out} C_{CO_2} + q_{H_2O}^{out} C_{H_2O} \tag{56}$$

$$C_c[CH_4 + N_2 + O_2] = q_{CH_4}^{out} C_{CH_4} + q_{N_2}^{in} C_{N_2} + q_{O_2}^{out} C_{O_2} \tag{57}$$

Following the previous section, these values can be calculated:

$C_r = C_{min}/C_{max}, q_{max} = C_{min}(T_{h,1} - T_{c,1})$

Assuming the SOFC configuration as parallel, the efficiency can be calculated:

$$\varepsilon = \frac{1 - e^{[-NTU(1+C_r)]}}{1 + C_r} \tag{58}$$

and $q = \varepsilon q_{max}$

Hence,

$$T_{h,2} = T_{h,1} - q/C_h \tag{59}$$

$$T_{c,2} = T_{c,1} + q/C_c \tag{60}$$

where,

- $T_{h,1}$=T:operating temperature
-

$$T_{c,1} = \frac{T_{CH_4}^{HX_2} q_{CH_4}^{out} + T_{O_2}^{HX_1} \left(q_{O_2}^{out} + q_{N_2}^{in}\right)}{q_{CH_4}^{out} + q_{O_2}^{out} + q_{N_2}^{in}} \tag{61}$$

- $T_{h,2} = T_{CO_2}^{SOFC} = T_{H_2O}^{SOFC}$ (reactants output temperature)
- $T_{c,2} = T_{SOFC-HX}$ (non-reactants output temperature)

The isothermal temperature of SOFC exhaust which will be fed into HX therefore can be calculated as:

$$T_{SOFC-exhaust} = \frac{T_{H_2O}^{SOFC} \left(q_{H_2O}^{out} + q_{CO_2}^{out}\right) + T_{SOFC-HX} \left(q_{CH_4}^{out} + q_{O_2}^{out} + q_{N_2}^{in}\right)}{q_{H_2O}^{out} + q_{CO_2}^{out} + q_{CH_4}^{out} + q_{O_2}^{out} + q_{N_2}^{in}} \tag{62}$$

3.3 Overall system energy efficiency comparison

As statement before, the aim to simulate HX included in SOFC simulation is to increase overall system energy efficiency. This section is going to point out the way to calculate the efficiency for SOFC model with HX versus one without HX. The efficiency of a chemical process must be evaluated differently than the conventional heat engine.

$$\eta_{withoutHX} = Electrical\ Energy_{per\ second} \times 100 / \Delta H [\%] \tag{63}$$

$$\eta_{withHX} = \frac{(\Delta H - Heat_{HXExhaust})_{per\ second} \times 100}{\Delta H} [\%] \tag{64}$$

$Heat_{HXExhaust}$ = Gas sensitive heat of HX exhaust.

4. Thermal dynamic model

4.1 Thermal balance model

Thermodynamics is the study of energy changing from one state to another. The predictions that can be made using thermodynamic equations are essential for understanding and modelling SOFC performance since SOFCs transform chemical energy into electrical energy. Basic thermodynamic concepts allow one to predict states of the SOFC system, such as potential, temperature, pressure, volume, and moles in a fuel cell.

The first few concepts relate to reacting systems in SOFC thermal balance analysis: absolute enthalpy, specific heat, entropy, and Gibbs free energy. The absolute enthalpy includes both chemical and sensible thermal energy. Chemical energy or the enthalpy of formation (h_f) is associated with the energy of the chemical bonds, and sensible thermal energy (Δh) is the enthalpy difference between the initial reactants and products of reaction. The next important property is specific heat, which is a measure of the amount of heat energy required to increase the temperature of a substance by 1^0C (or another temperature interval). Entropy (S) is another important concept, which is a measure of the quantity of heat that shows the possibility of conversion into work. Gibbs free energy (G) is the amount of useful work that can be obtained from an isothermal, isobaric system when the system changes from one set of steady-state conditions to another.

The temperature-voltage and voltage drop affect the life of material, the model should reflect this. The first step in determining the heat distribution in a fuel cell stack is to perform energy balances on the system. The total energy balance around the fuel cell is based upon the power produced, the fuel cell reactions, and the heat loss that occurs in a fuel cell. Heat losses include the convective heat transfer occurs between the solid surface and the gas streams, and the conductive heat transfer occurs in the solid and/or porous structures. The reactants, products, and electricity generated are the basic components to consider in modelling basic heat transfer in a fuel cell, as shown in Figure 9. Notice that in this model we have inserted the gas sensitive of the heat exchangers which will be explained in details in next section.

The general energy balance states that the enthalpy of the reactants and gas sensitive after HXs entering the fuel cell equals the enthalpy of the products leaving the cell plus the sum of the heat generated by the power output, and the rate of heat loss to the surroundings. The basic heat transfer calculations will aid in predicting the temperatures and heat in overall fuel cell stack and stack components.

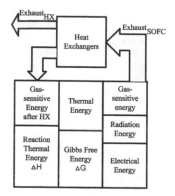

Fig. 9. Heat balance inside SOFC power unit with HX included

The parameters appear in thermal balance model are defined here.

- ΔH: Heat of reaction (Hess Energy) [J/mol] (Total energy produced by the reaction)
- ΔG: Thermal generation (Gibbs energy) [J/mol] (Electrical energy can retrieve a theoretical maximum)
- V: Cell terminal voltage [V]
- I_{fc}: Current [A]
- q_i: Species emissions [mol/s]
- R_i: Specific heat of water [J/mol.K]
- R_{FC}: Specific heat of the fuel cell system [J/K] (depends on manufacture)
- $\Delta H - IV$: Thermal Energy [J/mol]
- K_h: Coefficient of heat [W/K]
- T_0: Ambient temperature [K]
- T_{1i}, T_{2i}: The temperate at first and second state of species [K]
- T: Operating temperature [K]
- T_{ini}: The initial value Temperature [K]
- GSH_{HX}: Gas sensitive heat after HX [J/mol]
- GSH_{SOFC}: Gas sensitive heat after SOFC stack [J/mol]

Based on the diagram in Figure 9, the temperature equation and the flow of energy per unit time are as this representable expression.

$$GSH_{HX} + \Delta H - I_{fc}V = GSH_{SOFC} + R_{FC}\frac{dT}{dt} + K_h(T - T_0) \qquad (65)$$

IV is the output power, $K_h(T - T_0)$ is considered to be the energy dissipation. In the steady-state:

$$GSH_{HX} + \Delta H - I_{fc}V = GSH_{SOFC} + K_h(T - T_0) \qquad (66)$$

The dynamic operating temperature can be derived from Equation (2.65) shown in the following Equation.

$$T = T_{ini} + 1/R_{FC} \int_0^t \left\{ (GSH_{HX} + \Delta H - I_{fc}V) - K_h(T - T_0) - GSH_{SOFC} \right\} dt \qquad (67)$$

Thermal characteristics model is the model that uses the expression 67. A summary of this equation in Figure 10.

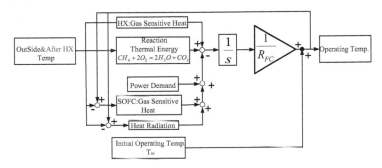

Fig. 10. Concept of thermal dynamic model

4.2 Heat factor

From the model shown in Figure 10, it is capable to calculate K_h by a independent model because this coefficient depend on the actual model and is independent with the change of operating temperature. It can be consider a constant with the chance of operating temperature. While Equation 66 is used, T can be considered as T_{ini}. Energy generated by fuel cells shown in Figure 10, and to assume that to change all the remaining heat energy to electrical energy extracted from the reaction of energy to representable fuel. Gibbs free energy is theoretically converted into electrical energy that can be used. Energy used in fuel cells is actually considered to be divided into three parts. And extract energy from the reaction energy as electricity output, the remaining energy is divided into heat radiation energy and heat energy is used to increase the temperature of the gas. From the Equation 65, the stack heat loss coefficient is obtained by dividing the amount of temperature change in the heat of the stack.

$$K_h = \frac{GSH_{HX} + \Delta H - I_{fc}V - GSH_{SOFC}}{T_{ini} - T_0} \qquad (68)$$

The conceptual model of above Equation is shown in Figure 11 which is derived by calculating the coefficient of heat energy from the stack number of moles of each gas equivalent reaction.

5. Simulation results of SOFC model implemented in Matlab/Simulink

The presented work is an attempt to model a SOFC system for DG applications. The aim of authors is to develop an efficient tool in Matlab-Simulink, which could simulate a SOFC system with sufficient accuracy. The reasons to use SIMULINK, are that the Matlab package

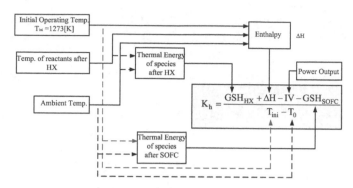

Fig. 11. Concept of heat loss coefficient calculation model

is commonly used among academic institutions and a graphical user interface with the high level of capabilities. Figure 12 shows the structure of the transient model.

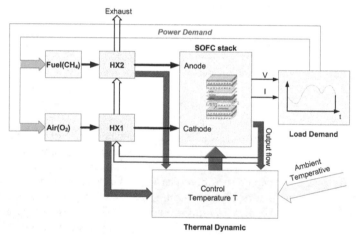

Fig. 12. Outline of single SOFC model

This SOFC model has simulated the internal reforming cell type, which generates hydrogen from methane by the high temperature of the cell. And its target is expected to follow changes in output load (50 [s] output at 70 [kW] to 100 [kW]). The transient analysis of the output is a directive issued. And the model also focuses on operating temperature control over long time (10.000s). The switch which has been set up in the center can simulate both if not considering the heat change for comparison. Each of the major SOFC components is be built as a complex sub-model (the load following control, air supply, fuel supply, the partial pressures, the voltage generator, the voltage drop, the temperature property, the heat exchangers, the efficiency calculation).

In order to carry out the SOFC model, it has to calculate partial pressures of 4 species and the chemical reaction for the SOFC terminal output voltage. The value of orders issued to meet the demand by the inverter power supply at the demand side by repeating this calculation, and numerical analysis. A feedback system has to be inserted for control the load following ability. The ac real power injection into the utility grid is considered to be the reference power for the fuel cell. The stack voltage and the reference power are used to determine the reference current which in turn is used to determine the fuel cell stack current (fuel and air supply). The fuel/air flow is proportional to the stack current. Throughout calculation system, the heat change affection is considered.

5.1 Load following ability

In this study, SOFC 70 [kW] power unit operating at 1273 [K] temperature works at normally stable load, 70 [kW] and the output command 100 [kW] is suddenly required for study of the response of SOFC. Observing SOFC load following ability, we need to consider the change of operating temperature.

Figure 13(a) is diagram of the output of current, voltage, power and operating temperature in short period test time, 200 [s]. This case considers robust following load change in small time scale. The output power takes 30 seconds to follow the increasing of load. That is equivalent to 1 [kW/s] velocity. The slow response of the fuel cell is due to the slow and gradual change in the fuel flow and the chemical reaction which is proportional to the stack current.

Figure 13(b) is diagram of the output of current, voltage, power and operating temperature in long period test time, 10000 [s]. This case considers the operating temperature. The temperature respond velocity with increase of load is 10 [K]/100 [s]. The final stable operating temperature is about 1050 [K] obtained at 2000s instance while the voltage and current output is about 437 [V], 230 [A].

Volt-amp characteristics of SOFC: The number of cells is taken to be 384 and the output voltage is 430 [V] which decreases as the load current increases. The drop is fairly linear in the middle region, known as region of Ohmic polarization. This is the operating region for the fuel cell.

5.2 Heat exchangers

5.2.1 Comparison between SOFC with and without heat exchangers

The simulation results are showed in Figure 14(a). Once we used HXs, according to the compared simulation results, the operating temperature be reduced by about 50[K].

5.2.2 Comparison between counter and parallel heat exchangers

The clear results are pointed out by Figure 14(b). The exhaust temperature into atmosphere of Counter-HX is lower than that of Parallel HX mean the higher energy efficiency. However, the operating temperature of SOFC in case using Counter-HX is higher than using Parallel-HX. The higher operating temperature makes stack materials work in severer condition. Based on this conclusion, the manufactures will decide which configuration is suitable for their real model.

5.2.3 Efficiency comparison

Figure 15(a) below shows the result that the efficiencies will change versus the times. The figure points out that the efficiencies change with the SOFC operation status within small range and when using HXs, the energy efficiencies are much higher.

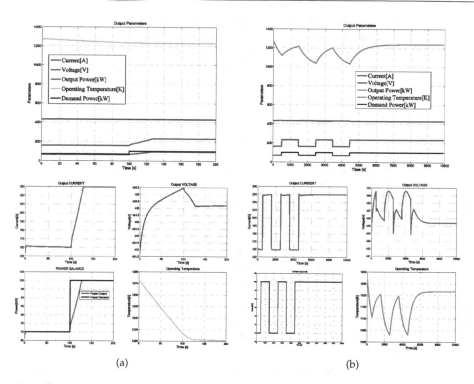

(a)   (b)

Fig. 13. a. Dynamics model in small timescale - 200[s]; b. Dynamics model in large timescale - 10000[s]

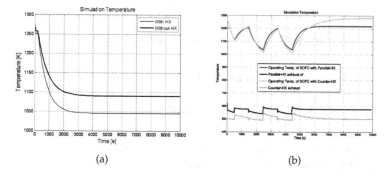

(a)   (b)

Fig. 14. a. Heat temperature comparison between SOFC with and without HX; b. HX operating temperature consideration between Counter and parallel Configuration

5.3 Operating temperature control by excess air

This section mentioned the study of temperature control in the SOFC operation. Rapid changes in heat in the fuel cell will lead to the deterioration of the material for the cell, SOFC it is important to properly maintain the temperature inside the stack. The operating temperature control method by using the excess air (O_2+N_2) into the fuel cell is shown in Figure 15(b).

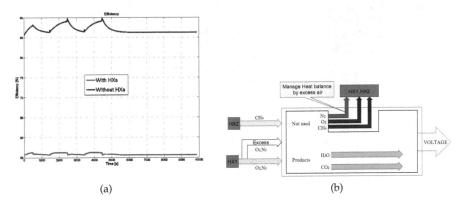

(a) (b)

Fig. 15. a. Efficiency comparison result; b. Concept of control temperature

Excess air is sent to take sensible heat of hot air, stabilize the temperature. The amount of excess air used is determined by the actual temperature which is adjusted by the feedback control.

The excess air bases on the required air for methane reactions, $q_{O_2}^{in}$. The amended air supply target is controlled by detecting the difference between the operating temperature and initial temperature 1273 [K]. The correction coefficient is determined by manufactures. This correction expression is shown in Equation 69. Figure 15(b) shows the model to control input air for maintaining operating temperature by using Equation 69.

$$q_{O_2-ex}^{in} = q_{O_2}^{in}[1 + K_{air}(\frac{T - 1273}{1273})]$$ (69)

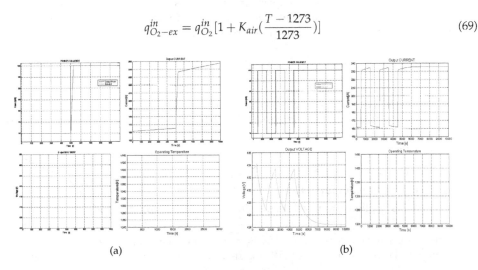

(a) (b)

Fig. 16. a. Without control of temperature - 1000[s]; b. Without control of temperature - 10000[s]

Following is the evaluation for using above excess air method. The output power at the various stages makes change in output voltage, current, operating temperature without excess

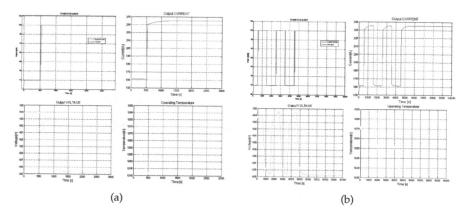

(a)   (b)

Fig. 17. a. Consideration control of temperature - 1000[s]; b. Consideration control of temperature - 10000[s]

air in Figure 16(a), 16(b). In Figure 55 simulation time is 1000 [s] with changing demand for one time, and in Figure 56, 10000 [s] with changing demand for several times are shown. These pictures pay attention to rapidly changing temperature and the SOFC stack material degradation will be suffered which reduces life time of SOFC stack.

The extra injection of air fed to control the temperature is shown in Figure 17(a), 17(b). Figure 17(a) expresses that the operating temperatures variation range is smaller over time. This temperature variation reducing can be expected to prevent the short fuel cell stack life because of the rapid changes in operating temperature. From this result, one idea emerges that if we fed SOFC stack with as much air flow rate as it needs to make the variation range smallest. This is actually impractical because of compressor configuration as well as the electrode pressure.

6. Conclusion

A dynamic model of SOFC power unit was developed in Simulink. The load change was subjected to a step change in the reference real power from 70 to 100 [kW]. The characteristics of the fuel cell (voltage, current and power) have a slower gradual change at the instant of step changes. Some goals of this chapter include:

1. Calculate heat balance inside SOFC power unit that effects on the operating temperature and therefore to the output voltage. The N_2 gas is also involved into consideration.
2. This model is applied for internal reforming that uses natural gas (CH_4) as a direct fuel.
3. Add the heat exchangers into SOFC power system and calculate the gas flows temperature attaching to heat balance to compact SOFC model. Evaluate the different heat characteristic of two popular configurations of HX, and therefore increasing system energy efficiency.
4. Evaluate the load following ability of SOFC power unit by using feedback control the fuel and air flow which respond to the load change.
5. Control the operating temperature by excess air.

7. References

A.C. Burt, R.S. Gemmen, A. S. (2004). A numerical study of cell-to-cell variations in a sofc stack, *Journal of Power Sources* Vol.126(76): 76–87.

Achenbach & Elmar (1995). Response of a solid oxide fuel cell to load change, *Journal of Power Sources* Vol.57: 105–109.

Ali Volkan Akkaya, B. S. & Erdem, H. H. (2009). Thermodynamic model for exegetic performance of a tubular sofc module, *Renewable Energy* Vol.1: 1–8.

Caisheng Wang, M. H. N. (2007). A physically based dynamic model for solid oxide fuel cells, *IEEE Transaction on Energy Conversion* Vol.22(4): 887–897.

CaishengWang, M.HashemNehrir, S. R. S. (2005). Dynamic models and model validation for pem fuel cells using electrical circuits, *IEEE Transaction on Energy Conversion* Vol.20(2): 442–451.

D. Sanchez, R. Chacartegui A. Mun, T. (2008). On the effect of methane internal reforming modeling in solid oxide fuel cells, *Journal of Hydrogen Energy* Vol.33: 1834–1844.

David L. Damm, A. G. F. (2005). Radiation heat transfer in sofc materials and components, *Journal of Power Sources* Vol.143: 158–165.

Graham M. Goldin, Huayang Zhu, R. D. B. S. A. B. (2009). Multidimensional flow, thermal, and chemical behavior in solid-oxide fuel cell button cells, *Journal of Power Sources* Vol.187: 123–135.

J. Padulles, G.W. Ault, J. M. (2000). A numerical study of cell-to-cell variations in a sofc stack, *Journal of Power Sources* Vol.86: 495–500.

James Larminie, A. D. (2003). *Fuel Cell Systems Explained, 2nd ed.*, Wiley, England.

Kourosh Sedghisigarchi, A. F. (2004). Dynamic and transient analysis of power distribution systems with fuel cells-part i:fuel-cell dynamic mode, *IEEE Transaction on Energy Conversion* Vol.19(2): 423–428.

Li, P.-W. & Chyu, M. K. (2003). Simulation of the chemicalsimulation of the chemical/electrochemical reactions and heat/mass transfer for a tubular sofc in a stack, *Journal of Power Sources* Vol.124: 487–498.

M. Uzunoglu, M. (2006). Dynamic modeling, design, and simulation of a combined pem fuel cell and ultracapacitor system for stand-alone residential applications, *IEEE Transaction on Energy Conversion* Vol.21(3): 767.

Mitsunori Iwata, Takeshi Hikosaka, M. M. T. I. K. I. K. O. Y. E. Y. S. S. N. (2000). Performance analysis of planar-type unit sofc considering current and temperature distributions, *Solid State Ionics* Vol.132: 297–308.

M.Y. El-Sharkh, A. Rahman, M. A. P. B. A. S. T. T. (2004). A dynamic model for a stand-alone pem fuel cell power plant for residential applications, *Journal of Power Sources* Vol.138: 199–204.

N. Lu, Q. Li, X. S. M. K. (2006). The modeling of a standalone solid-oxide fuel cell auxiliary power unit, *Journal of Power Sources* Vol.161: 938–948.

of Energy, U. D. (2004). *Fuel Cell Handbook, 7th ed.*, EG and G Technical Services, Inc., Morgantown, West Virginia, USA.

P. Piroonlerkgul, W. Kiatkittipong, A. A. A. S. W. W. N. L. A. A. S. A. (2009). Integration of solid oxide fuel cell and palladium membranereactor: Technical and economic analysis, *International journal of hydrogen energy* Vol.34(9): 3894–3907.

P.R. Pathapati, X. Xue, J. T. (2005). A new dynamic model for predicting transient phenomena in a pem fuel cell system, *Journal of Renewable Energy* Vol.30: 1–22.

S. Campanari, P. I. (2004). Definition and sensitivity analysis of a finite volume sofc model for a tubular cell geometry, *Journal of Power Sources* Vol.132: 113–126.

S.H. Chan, K.A. Khor, Z. X. (2001). A complete polarization model of a solid oxide fuel cell and its sensitivity to the change of cell component thickness, *Journal of Power Sources* Vol.93: 130–140.

S.H.Chan, C.F.Low, O. (2002). Energy and energy analysis of simple solid-oxide fuel-cell power system, *Journal of Power Sources* Vol.103: 188–200.

Susumu Nagata, Akihiko Momma, T. K. Y. K. (2001). Numerical analysis of output characteristics of tubular sofc with internal reformer, *Journal of Power Sources* Vol.101: 60–71.

Tadashi Gengo, Nagao Hisatome, Y. A. Y. K. T. K. K. K. (2007). Progressing steadily, development of high-efficiency sofc combined cycle system, *Mitsubishi Heavy Industries, Ltd.Technical Review* Vol.44(1): 1–5.

Takanobu Shimada, Akihiko Momma, K. T. T. K. (2009). Numerical analysis of electrical power generation and internal reforming characteristics in seal-less disk-type solid oxide fuel cells, *Journal of Power Sources* Vol.187: 8–18.

Tomoyuki Ota, Michihisa Koyama, C.-j. W. K. Y. H. T. (2003). Object-based modeling of sofc system: dynamic behavior of micro-tube sofc, *Journal of Power Sources* Vol.118: 430–439.

Wang, C. & Nehrir, M. H. (2007). Load transient mitigation for stand-alone fuel cell power generation systems, *IEEE Transaction on Energy Conversion* Vol.22(4): 864–872.

Xiongwen Zhang, Guojun Li, J. L. Z. F. (2007). Numerical study on electric characteristics of solid oxide fuel cells, *Energy Conversion and Management* Vol.48: 977–989.

Permissions

The contributors of this book come from diverse backgrounds, making this book a truly international effort. This book will bring forth new frontiers with its revolutionizing research information and detailed analysis of the nascent developments around the world.

We would like to thank Dr. Hamid A. Al-Megren, for lending his expertise to make the book truly unique. He has played a crucial role in the development of this book. Without his invaluable contribution this book wouldn't have been possible. He has made vital efforts to compile up to date information on the varied aspects of this subject to make this book a valuable addition to the collection of many professionals and students.

This book was conceptualized with the vision of imparting up-to-date information and advanced data in this field. To ensure the same, a matchless editorial board was set up. Every individual on the board went through rigorous rounds of assessment to prove their worth. After which they invested a large part of their time researching and compiling the most relevant data for our readers. Conferences and sessions were held from time to time between the editorial board and the contributing authors to present the data in the most comprehensible form. The editorial team has worked tirelessly to provide valuable and valid information to help people across the globe.

Every chapter published in this book has been scrutinized by our experts. Their significance has been extensively debated. The topics covered herein carry significant findings which will fuel the growth of the discipline. They may even be implemented as practical applications or may be referred to as a beginning point for another development. Chapters in this book were first published by InTech; hereby published with permission under the Creative Commons Attribution License or equivalent.

The editorial board has been involved in producing this book since its inception. They have spent rigorous hours researching and exploring the diverse topics which have resulted in the successful publishing of this book. They have passed on their knowledge of decades through this book. To expedite this challenging task, the publisher supported the team at every step. A small team of assistant editors was also appointed to further simplify the editing procedure and attain best results for the readers.

Our editorial team has been hand-picked from every corner of the world. Their multi-ethnicity adds dynamic inputs to the discussions which result in innovative outcomes. These outcomes are then further discussed with the researchers and contributors who give their valuable feedback and opinion regarding the same. The feedback is then collaborated with the researches and they are edited in a comprehensive manner to aid the understanding of the subject.

Apart from the editorial board, the designing team has also invested a significant amount of their time in understanding the subject and creating the most relevant covers. They scrutinized every image to scout for the most suitable representation of the subject and create an appropriate cover for the book.

The publishing team has been involved in this book since its early stages. They were actively engaged in every process, be it collecting the data, connecting with the contributors or procuring relevant information. The team has been an ardent support to the editorial, designing and production team. Their endless efforts to recruit the best for this project, has resulted in the accomplishment of this book. They are a veteran in the field of academics and their pool of knowledge is as vast as their experience in printing. Their expertise and guidance has proved useful at every step. Their uncompromising quality standards have made this book an exceptional effort. Their encouragement from time to time has been an inspiration for everyone.

The publisher and the editorial board hope that this book will prove to be a valuable piece of knowledge for researchers, students, practitioners and scholars across the globe.

List of Contributors

Joseph Essandoh-Yeddu
University of Cape Coast, Energy Commission, Ghana

Blanca E. García-Flores and Fernando García-Sánchez
Laboratory of Thermodynamics, Research Program in Molecular Engineering, Mexican Petroleum Institute, Mexico, D.F., Mexico

Daimler N. Justo-García
Department of Chemical and Petroleum Engineering, ESIQIE, National Polytechnic Institute, Mexico, D.F., Mexico

Roumiana P. Stateva
Institute of Chemical Engineering, Bulgarian Academy of Sciences, Sofia, Bulgaria

Jolanta Szoplik
West Pomeranian University of Technology, Szczecin, Poland

Tiancun Xiao
Inorganic Chemistry Laboratory, Oxford University, UK
Guangzhou Boxenergy Technology Ltd, Guangzhou, PR China

Hamid Al-Megeren
Petrochemical Research Institute, King Abdulaziz City for Science and Technology, Riyadh, Saudi Arabia

Victor M. Maslennikov, Vyacheslav M. Batenin, Yury A. Vyskubenko and Victor Ja. Shterenberg
Joint Institute for High Temperatures of the Russian Academy of Sciences, Russian Federation

Ranajit K. Saha
Ex-HOD, Department of Chemical Engg., Indian Institute of Technology, Kharagpur, India

Pankaj Mathure
Epoxy Division, Aditya Birla Chemicals Ltd., Thailand

Anand V. Patwardhan
Department of Chemical Engg., Institute of Chemical Technology, Mumbai, India

Małgorzata Wilk and Aneta Magdziarz
AGH University of Science and Technology, Krakow, Poland

Rosli Abu Bakar, K. Kadirgama, M.M. Rahman and K.V. Sharma
Faculty of Mechanical Engineering, University Malaysia Pahang, Malaysia

Semin
Department of Marine Engineering, Institut Teknologi Sepuluh Nopember, Surabaya, Indonesia

Norberto Pérez Rodríguez, Erik Rosado Tamariz, Alfonso Campos Amezcua and Rafael García Illescas
Instituto de Investigaciones Eléctricas (IIE), Mexico

Nguyen Duc Tuyen and Goro Fujita
Shibaura Institute of Technology, Japan

9 781632 403711